ENERGY SCIENCE, ENGINEERING AND TECHNOLOGY

UNCONVENTIONAL OIL AND SHALE GAS

GROWTH, EXTRACTION, AND WATER MANAGEMENT ISSUES

ENERGY SCIENCE, ENGINEERING AND TECHNOLOGY

Additional books in this series can be found on Nova's website under the Series tab.

Additional e-books in this series can be found on Nova's website under the e-book tab.

ENERGY SCIENCE, ENGINEERING AND TECHNOLOGY

UNCONVENTIONAL OIL AND SHALE GAS

GROWTH, EXTRACTION, AND WATER MANAGEMENT ISSUES

AMBER L. TUFT
EDITOR

nova publishers
New York

Library of Congress Cataloging-in-Publication Data

ISBN: 978-1-63482-130-8

Published by Nova Science Publishers, Inc. † New York

CONTENTS

PREFACE

This book focuses on the growth in U.S. oil and natural gas production driven primarily by tight oil formations and shale gas formations. It reviews selected federal environmental regulatory and research initiatives related to unconventional oil and gas extraction, including the Bureau of Land Management (BLM) proposed hydraulic fracturing rule and Environmental Protection Agency (EPA) actions. This book also provides a technological assessment of existing and emerging water procurement and management practices in shale energy producing regions of the United States.

Chapter 1 – The United States has seen resurgence in petroleum production, mainly driven by technology improvements—especially hydraulic fracturing and directional drilling—developed for natural gas production from shale formations. Application of these technologies enabled natural gas to be economically produced from shale and other unconventional formations, and contributed to the United States becoming the world's largest natural gas producer in 2009. Use of these technologies has also contributed to the rise in U.S. oil production over the last few years. In 2009, annual oil production increased over 2008, the first annual rise since 1991, and has continued to increase each year since. Between January 2008 and May 2014, U.S. monthly crude oil production rose by 3.2 million barrels per day, with about 85% of the increase coming from shale and related tight oil formations in Texas and North Dakota. Other tight oil plays are also being developed, helping raise the prospect of energy independence, especially for North America.

The rapid expansion of tight oil and shale gas extraction using high-volume hydraulic fracturing has raised concerns about its potential environmental and health impacts. These concerns include potential direct impacts to groundwater and surface water quality, water supplies, and air quality. In addition, some have raised concerns about potential long-term and indirect impacts from reliance on fossil fuels and resulting greenhouse gas emissions and influence on broader energy economics. This report focuses mainly on actions related to controlling potential direct impacts.

States are the primary regulators of oil and gas production on non-federal lands. State laws and regulations governing oil and gas production have been evolving across the states in response to changes in production practices as producers have expanded into tight oil, shale gas, and other unconventional hydrocarbon formations. However, state rules vary considerably, leading to calls for more federal oversight of unconventional oil and gas extraction activities, and hydraulic fracturing specifically.

Although provisions of several federal environmental laws can apply to certain activities related to oil and gas production, proposals to expand federal regulation in this area have been highly controversial. Some advocates of a larger federal role point to a wide range of differences among state regulatory regimes, and argue that a national framework is needed to ensure a consistent minimum level of protection for surface and groundwater resources, and air quality. Others argue against more federal involvement, and point to the long-established state oil and natural gas regulatory programs, regional differences in geology and water resources, and concern over regulatory redundancy.

The federal role in regulating oil and gas extraction activities—and hydraulic fracturing, in particular—has been the subject of considerable debate and legislative proposals for several years, but legislation has not been enacted. While congressional debate has continued, the Administration has pursued a number of regulatory initiatives related to unconventional oil and gas development under existing statutory authorities.

This report focuses on the growth in U.S. oil and natural gas production driven primarily by tight oil formations and shale gas formations. It also reviews selected federal environmental regulatory and research initiatives related to unconventional oil and gas extraction, including the Bureau of Land Management (BLM) proposed hydraulic fracturing rule and Environmental Protection Agency (EPA) actions.

Chapter 2 - Shale oil and gas (collectively referred to as shale energy), long considered "unconventional" hydrocarbon resources, are now being developed rapidly. Economic extraction of shale energy resources typically relies on the use of hydraulic fracturing. This technique often requires significant amounts of freshwater, and fracturing flowback and related wastewaters must be recycled or disposed of after a well is completed. While shale energy presents a significant energy resource, its development has the potential to pose risks to water availability and water quality.

This report provides a technological assessment of existing and emerging water procurement and management practices in shale energy-producing regions of the United States. The intersection of evolving technology, growing environmental concerns, demand for new sources of hydrocarbon energy, and the potential national interests in developing shale oil and gas resources provides the context for this study. Congressional attention has been focused on two key aspects of the issue: shale energy as a growing U.S. energy source, and environmental concerns associated with the development of these resources.

Water for shale energy projects is used most intensely in the fracturing portion of a well's life cycle. Under current practices, fracturing typically is a water-dependent activity, often requiring between a few million and 10 million gallons of water per fractured horizontal well. This water demand often is concentrated geographically and temporally during the development of a particular shale formation. Production activities and management and treatment of the wastewater produced during shale energy production (including flowback from fracturing and water produced from source formations) have raised concerns over the potential contamination of groundwater and surface water and induced seismicity associated with wastewater injection wells.

Water resource issues may pose constraints on the future development of domestic shale oil and gas. Potential negative effects from shale energy extraction—particularly effects associated with hydraulic fracturing and wastewater management—have prompted state and regional regulatory actions to protect water supplies. Future congressional and executive branch actions may influence development of shale oil and shale gas on federal lands and elsewhere through additional regulatory oversight or other policy actions. At the same time, advances in shale energy extraction and wastewater management techniques may reduce some development impacts.

The pace of technological change in water sourcing and water management in the shale energy sector is rapid, but uneven. Trends in water management have generally been influenced by local disposal costs, regulations, and geologic conditions rather than by water scarcity alone. Emerging technologies and practices in water resources management can be divided into those that seek to reduce the amount of consumptive freshwater utilization in the drilling and completion process, and those that seek to lower the costs and/or minimize the potential for negative environmental impacts associated with wastewater management.

Water management issues are relevant to the entire life cycle of shale energy development, because fluids will continue to be produced even after a well is drilled, fractured, and producing oil and/or natural gas. Research that views the shale energy production process in a life-cycle and materials-flow context may facilitate the identification of technologies and processes that can mitigate potential impacts along different stages of shale energy development.

Chapter 3 – Hydraulic fracturing is a technique developed initially to stimulate oil production from wells in declining oil reservoirs. With technological advances, hydraulic fracturing is now widely used to initiate oil and gas production in unconventional (low-permeability) oil and gas formations that were previously uneconomical to produce. This process now is used in more than 90% of new oil and gas wells and in many existing wells to stimulate production. Hydraulic fracturing is done after a well is drilled, and involves injecting large volumes of water, sand (or other propping agent), and specialized chemicals under enough pressure to fracture the formations holding the oil or gas. The sand or other proppant holds the fractures open to allow the oil or gas to flow freely out of the formation and into a production well. Its application, in combination with horizontal drilling, for production of natural gas (methane) from tight gas sands, unconventional shale formations, and coal beds, has resulted in the marked expansion of estimated U.S. natural gas reserves in recent years. Similarly, hydraulic fracturing is enabling the development of tight oil resources, such as the Bakken and Eagle Ford formations. The rapid growth in the use of fracturing has raised concerns over its potential impacts on groundwater and drinking water sources, and has led to calls for more state and/or federal oversight of this activity.

Historically, the Environmental Protection Agency (EPA) had not regulated the underground injection of fluids for hydraulic fracturing of oil or gas production wells. In 1997, the U.S. Court of Appeals for the 11th Circuit ruled that fracturing for coalbed methane (CBM) production in Alabama constituted underground injection and must be regulated under the Safe Drinking Water Act (SDWA). This ruling led EPA to study the risk that hydraulic fracturing for CBM production might pose to drinking water sources. In 2004, EPA reported that the risk was small, except where diesel was used, and that national regulation was not needed. However, to address regulatory uncertainty the ruling created, the Energy Policy Act of 2005 (EPAct 2005) revised the SDWA term "underground injection" to explicitly exclude the injection of fluids and propping agents (except diesel fuels) used for hydraulic fracturing purposes. Thus EPA lacks authority under the SDWA to regulate hydraulic fracturing, except where diesel fuels are used. In February 2014, EPA issued final permitting guidance for hydraulic fracturing operations using diesel fuels.

As the use of the process has grown, some in Congress would like to revisit the 2005 statutory exclusion. Legislation to revise the act's definition of underground injection to explicitly include hydraulic fracturing has been offered in recent years, but not enacted. A variety of hydraulic fracturing bills are pending in the 113th Congress. In EPA's FY2010 appropriations act, Congress urged the agency to study the relationship between hydraulic fracturing and drinking water quality. In 2012, EPA issued a research progress report. The agency expects to issue a final report in 2016.

This report reviews past and proposed treatment of hydraulic fracturing under the SDWA, the principal federal statute for regulating the underground injection of fluids to protect groundwater sources of drinking water. It reviews current SDWA provisions for regulating underground injection activities, and discusses some possible implications of the enactment of legislation authorizing EPA to regulate hydraulic fracturing (beyond diesel) under this statute. The report also reviews legislative proposals concerning the regulation of hydraulic fracturing under the SDWA.

In: Unconventional Oil and Shale Gas
Editor: Amber L. Tuft
ISBN: 978-1-63482-130-8

Chapter 1

AN OVERVIEW OF UNCONVENTIONAL OIL AND NATURAL GAS: RESOURCES AND FEDERAL ACTIONS*

Michael Ratner and Mary Tiemann

SUMMARY

The United States has seen resurgence in petroleum production, mainly driven by technology improvements—especially hydraulic fracturing and directional drilling—developed for natural gas production from shale formations. Application of these technologies enabled natural gas to be economically produced from shale and other unconventional formations, and contributed to the United States becoming the world's largest natural gas producer in 2009. Use of these technologies has also contributed to the rise in U.S. oil production over the last few years. In 2009, annual oil production increased over 2008, the first annual rise since 1991, and has continued to increase each year since. Between January 2008 and May 2014, U.S. monthly crude oil production rose by 3.2 million barrels per day, with about 85% of the increase coming from shale and related tight oil formations in Texas and North Dakota. Other tight oil plays are also being developed, helping raise the prospect of energy independence, especially for North America.

---

* This is an edited, reformatted and augmented version of a Congressional Research Service publication R43148, prepared for Members and Committees of Congress dated November 21, 2014.

The rapid expansion of tight oil and shale gas extraction using high-volume hydraulic fracturing has raised concerns about its potential environmental and health impacts. These concerns include potential direct impacts to groundwater and surface water quality, water supplies, and air quality. In addition, some have raised concerns about potential long-term and indirect impacts from reliance on fossil fuels and resulting greenhouse gas emissions and influence on broader energy economics. This report focuses mainly on actions related to controlling potential direct impacts.

States are the primary regulators of oil and gas production on non-federal lands. State laws and regulations governing oil and gas production have been evolving across the states in response to changes in production practices as producers have expanded into tight oil, shale gas, and other unconventional hydrocarbon formations. However, state rules vary considerably, leading to calls for more federal oversight of unconventional oil and gas extraction activities, and hydraulic fracturing specifically.

Although provisions of several federal environmental laws can apply to certain activities related to oil and gas production, proposals to expand federal regulation in this area have been highly controversial. Some advocates of a larger federal role point to a wide range of differences among state regulatory regimes, and argue that a national framework is needed to ensure a consistent minimum level of protection for surface and groundwater resources, and air quality. Others argue against more federal involvement, and point to the long-established state oil and natural gas regulatory programs, regional differences in geology and water resources, and concern over regulatory redundancy.

The federal role in regulating oil and gas extraction activities—and hydraulic fracturing, in particular—has been the subject of considerable debate and legislative proposals for several years, but legislation has not been enacted. While congressional debate has continued, the Administration has pursued a number of regulatory initiatives related to unconventional oil and gas development under existing statutory authorities.

This report focuses on the growth in U.S. oil and natural gas production driven primarily by tight oil formations and shale gas formations. It also reviews selected federal environmental regulatory and research initiatives related to unconventional oil and gas extraction, including the Bureau of Land Management (BLM) proposed hydraulic fracturing rule and Environmental Protection Agency (EPA) actions.

INTRODUCTION: CHANGE IS AFOOT

In the past, the oil and natural gas industry considered resources locked in tight, impermeable formations such as shale uneconomical to produce. Advances in directional well drilling and reservoir stimulation, however, have dramatically changed this perspective. It is production from these unconventional formations that has changed the U.S. energy posture and global energy markets.

U.S. oil and natural gas production is on the rise, primarily driven by resources from tight formations. The techniques developed to produce shale gas—directional drilling and hydraulic fracturing[1]—have migrated to the oil sector. The United States is the third-largest oil producer in the world, but also the fastest-growing producer. The United States surpassed Russia in 2009 as the world's largest natural gas producer. Production from tight formations is expected to make up a significant part of production of each commodity well into the future (see *Figure 1*).

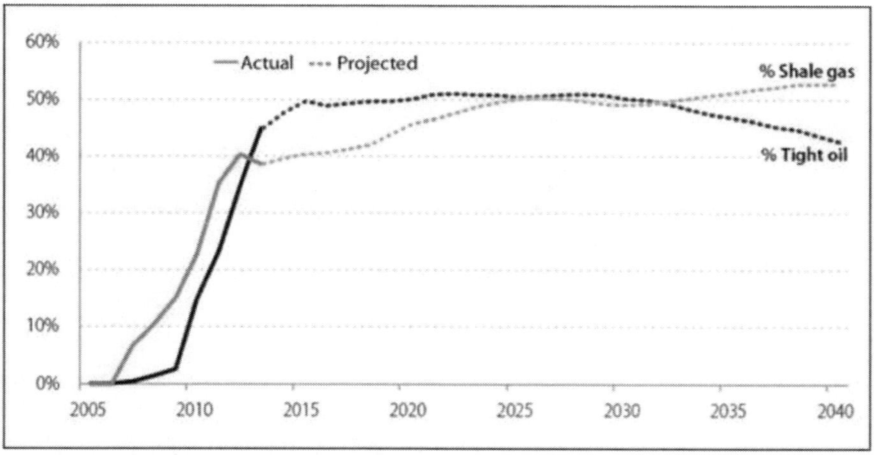

Source: U.S. Energy Information Administration, Annual Energy Outlook 2014, http://www.eia.gov/oiaf/aeo/ tablebrowser/ and other EIA data.

Note: Prior to 2007, the Energy Information Administration did not report tight oil and shale gas data.

Figure 1. Percentage of U.S. Oil and Natural Gas from Tight Oil and Shale Gas, (2005-2040).

This report focuses on the growth in U.S. oil and natural gas production driven primarily by tight oil formations and shale gas formations. It does not address other types of unconventional production such as coalbed methane or tight gas, as their contributions to overall U.S. production have not changed as dramatically as shale gas.[2] There has been continued congressional interest through the 113th Congress related to unconventional natural gas and oil production. In May 2013, the Senate Energy and Natural Resources Committee held three roundtable discussions on natural gas supply and use.[3] The House Energy and Commerce Committee's Subcommittee on Energy and Power held a hearing in June 2013 on U.S. energy abundance.[4]

GEOLOGY IS WHAT MAKES A RESOURCE UNCONVENTIONAL

Unconventional formations are fine-grained, organic-rich, sedimentary rocks—usually shales and similar rocks. The shales and rocks are both the source of and the reservoir for oil and natural gas, unlike conventional petroleum reservoirs. The Society of Petroleum Engineers describes "unconventional resources" as petroleum accumulations that are pervasive throughout a large area and that are not significantly affected by pressure exerted by water (hydrodynamic influences); they are also called "continuous-type deposits" or "tight formations." In contrast, conventional oil and natural gas deposits occur in porous and permeable sandstone and carbonate reservoirs. Under pressure exerted by water, the hydrocarbons migrated upward from organic sources until an impermeable cap-rock (such as shale) trapped it in the reservoir rock. Although the unconventional formations may be as porous as other sedimentary reservoir rocks, their extremely small pore sizes and lack of permeability make them relatively resistant to hydrocarbon flow. The lack of permeability means that the oil and gas typically remain in the source rock unless natural or artificial fractures occur.

PRICE DRIVES INDUSTRIAL INNOVATION

Historically, natural gas prices in the United States have been volatile. From 1995 to 1999 the spot price of natural gas averaged \$2.23 per million British thermal units (MBtu), but increased to an average price of \$4.68 per

MBtu, in nominal dollars, during the 2000-to-2004 period, an almost 110% rise. Prices hit a peak in December 2005 at $15.38 per MBtu, but remained relatively high through July 2008, as can be seen in *Figure 2.* Along with the rise in prices, U.S. net imports of natural gas also rose, increasing 32% between 1995 and 2000 and 41% between 1995 and 2007.

As U.S. prices and imports continued to trend up, industry undertook two competing solutions to meet the need for more natural gas—increased liquefied natural gas (LNG) imports and development of techniques to produce shale gas. The LNG import facilities were much higher-profile and were cited extensively in industry and popular press. Approximately 50 import projects were proposed, and eight were eventually constructed during the mid- to late 2000s, along with the recommissioning of older facilities.

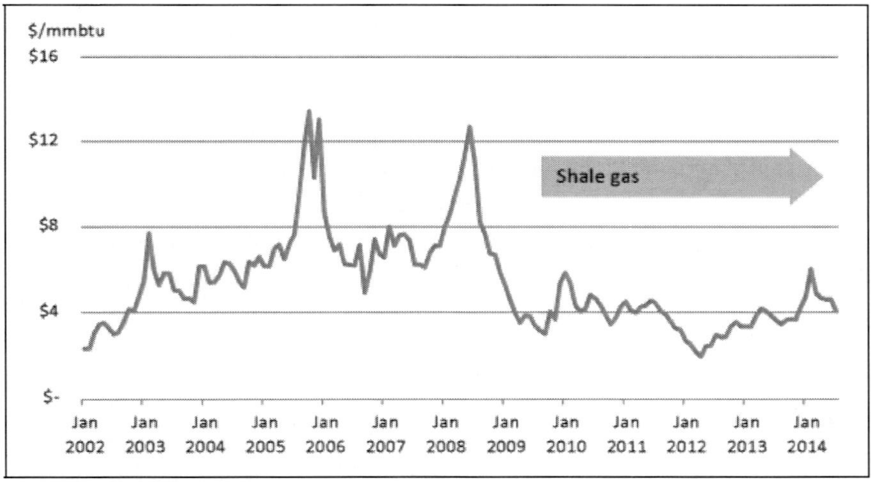

Source: U.S. Energy Information Administration, http://www.eia.gov/dnav/ng/hist/rngw hdM.htm.

Notes: Units = nominal dollars per million British thermal units (mmBtu). Data for 2014 are through July.

Figure 2. Monthly U.S. Natural Gas Prices, (2002-2014).

Although horizontal drilling and hydraulic fracturing have been industry techniques for some time, their application to shale gas formations is relatively new. Advances in directional drilling, particularly steerable down-hole motors, allowed drilling operators to better keep the well bore in the hydrocarbon-bearing shale formations. Well stimulation was also required, and improvements in hydraulic fracturing techniques, particularly multistage hydraulic fracturing and the ability to better control the fractures, contributed to making shale gas production a profitable venture.

In 2007, the Energy Information Administration (EIA) first recorded shale gas production, when it accounted for just 7% of U.S. natural gas production. In 2013, shale gas production accounted for almost 40% of U.S. production (see *Figure 1*), while almost all the LNG import terminals were idle and many applied to become export terminals.[5]

TECHNOLOGIES STIMULATE SHALE GAS PRODUCTION FIRST

The application of advances in directional drilling and hydraulic fracturing were first applied to shale gas formations, particularly as natural gas prices increased in the mid-2000s. Methane molecules and those of natural gas liquids (NGLs) are smaller than crude oil molecules and therefore tend to be more responsive to hydraulic fracturing. The success of shale gas development has driven U.S. natural gas production to increase almost every month on a year-on-year basis (see *Figure 3*) from 2008 through May 2014. The rise in shale gas development has also resulted in natural gas prices declining, as shown in *Figure 2*.

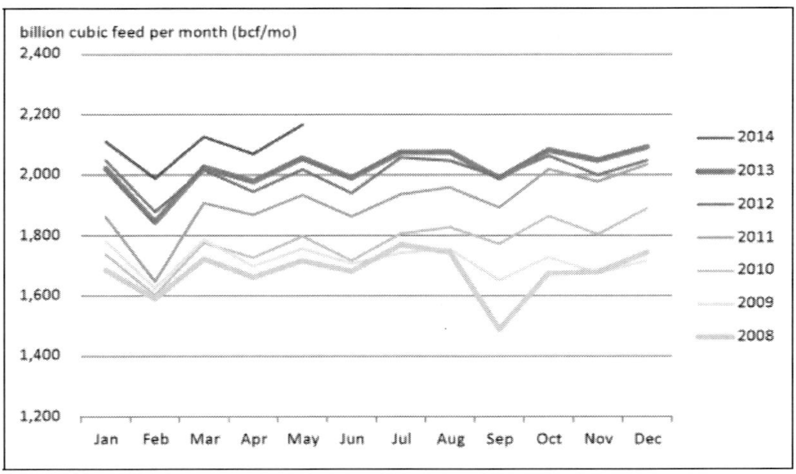

Source: U.S. Energy Information Administration, http://www.eia.gov/dnav/ng/ng_prod_sum_dcu_NUS_m.htm.

Figure 3. Monthly U.S. Natural Gas Production, (2008-2014).

The decline in prices and production in the latter half of 2008 was mainly the result of the economic downturn. However, as the economy picked up in 2009, natural gas resumed its upward production trajectory while prices stayed low. Overall U.S. natural gas production grew, as did the contribution from shale. The continued increase in production can be attributed, in part, to industry improvements in extracting more of the natural gas from the shale formations. Continued progress in hydraulic fracturing and directional drilling techniques has enabled companies to drive down production costs while increasing output.

Natural Gas Liquids: A Production Driver

Natural gas liquids (NGLs) have taken on a new prominence as shale gas production has increased and prices have fallen. As natural gas prices have stayed low, company interests have shifted away from dry natural gas production to more liquids-based production. NGL is a general term for all liquid products separated from natural gas at a gas processing plant, and includes ethane, propane, butane, and pentanes. When NGLs are present with methane, which is the primary component of natural gas, the natural gas is referred to as either "hot" or "wet" gas. Once the NGLs are removed from the methane, the natural gas is referred to as "dry" gas, which is what most consumers use.

Each NGL has its own market and its own value. As the price for dry gas has dropped because of the increase in supply and other reasons, such as the warm winter of 2011, the natural gas industry has turned its attention to producing in areas with more wet gas in order to bolster the value it receives (see *Figure 4*). Some companies have shifted their production portfolios to tight oil formations, such as the Bakken in North Dakota and Montana, to capitalize on the experience they gained in shale gas development. Historically, the individual NGL products have been priced against oil, except for ethane. As oil prices have remained higher since 2008 relative to natural gas, they have driven an increase of wet gas production. Because of its low price, dry gas is often treated as a "by-product" of wet gas and oil production.

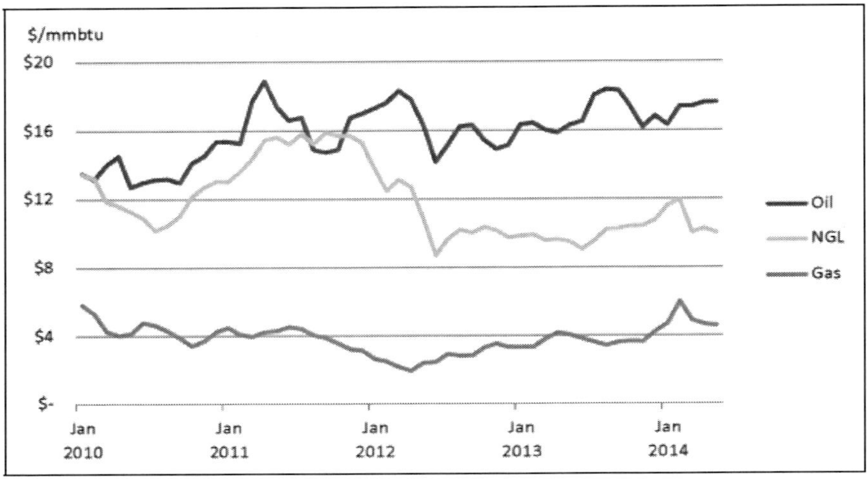

Source: U.S. Energy Information Administration.

Notes: According to EIA, the NGL composite price is derived from daily Bloomberg spot price data for natural gas liquids at Mont Belvieu, TX, weighted by gas processing plant production volumes of each product as reported on Form EIA-816, "Monthly Natural Gas Liquids Report." The mix of NGLs will vary by source, and the price will vary by the actual market for the product. The natural gas price is at Henry Hub, and the oil price is West Texas Intermediate (WTI). Units = nominal dollars per million British thermal units ($/mmBtu). Data for 2014 are through May.

Figure 4. Natural Gas, Oil, and NGL Prices, (2010-2014).

Increased Tight Oil Production Raises Independence Possibility

The prospect of U.S. energy independence is grounded in the production growth from tight oil formations such as the Bakken Formation in North Dakota and Montana, and the Eagle Ford Formation in Texas.[6] Relative to other fuels, the United States is more dependent upon imports for its oil requirements, still accounting for about 47% of consumption.[7] Canada is the largest supplier of U.S. oil imports, which is why energy independence is usually mentioned as North American energy independence.[8] The United States added almost 1 million barrels per day (b/d) of oil production between 2012 and 2013 (see *Figure 5*). U.S. oil production has reached levels not seen in more than a decade, but is almost 2 million b/d short of the highs in the

1970s. Since 2005, when crude oil imports reached a peak, they have dropped almost 2.4 million b/d, or 24%, through 2013.[9] Also since 2005, U.S. consumption of crude oil and petroleum products has been trending downward, contributing to the decrease in imports.

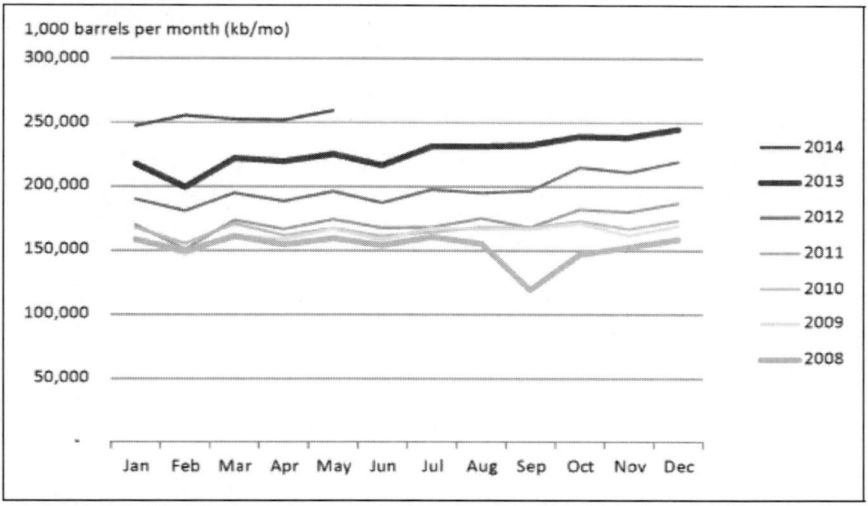

Source: U.S. Energy Information Administration, http://www.eia.gov/dnav/pet/hist/LeafHandler.ashx?n=PET&s=MCRFPUS1&f=M.

Figure 5. Monthly U.S. Oil Production, (2008-2014).

The continued shift of industry resources toward oil-rich production has prompted forecasts of continued growth. Domestic crude oil production is projected to rise through the end of the decade. The tremendous increases are primarily due to dramatic increases in production from the previously mentioned Bakken Formation in North Dakota and the Eagle Ford play in Texas, both tight oil formations.[10]

Environmental Concerns and Responses

As with other energy sources or fuel production, the development of unconventional oil and gas resources can pose both environmental risks and net benefits, some direct and others indirect. Potential direct risks may include impacts to groundwater and surface water quality, public and private water supplies, and air quality. In addition, some have raised concerns about potential long-term and indirect impacts from reliance on fossil fuels and resulting greenhouse gas emissions and influence on broader energy economics. On the other hand, natural gas is seen by many as a "bridge" fuel that can provide more energy per unit of greenhouse gas produced than some alternatives (e.g., coal), and has only recently been produced in sufficient quantity and at low enough prices to provide a viable alternative fuel that is widely regarded as relatively cleaner-burning (i.e., no mercury or sulfur emissions and substantially lower emissions of nitrous oxides (NOx) and carbon dioxide (CO_2) per Btu of energy produced compared to coal). This report focuses primarily on measures to address potential direct impacts.

Among the variety of potential direct environmental impacts, many may be mitigated with appropriate safeguards, existing technology, and best practices. For example, management of wastewater associated with increased unconventional oil and gas production activity has in some cases placed a strain on water resources, and on wastewater treatment plants that were not designed to remove salts and other contaminants from hydraulic fracturing flowback and produced water, and these impacts can be mitigated by investing in additional control technologies.

Water quality issues have received much attention, and of these, the potential risks associated with well stimulation by hydraulic fracturing have been at the forefront. Complaints of contaminated well water have emerged in some areas where unconventional oil and gas development has occurred, although regulators have not reported a direct connection between hydraulic fracturing of shale formations at depth and groundwater contamination. In shale formations, the vertical distance separating the target zone from usable aquifers generally is much greater than the length of the fractures induced during hydraulic fracturing. Thousands of feet of rock layers typically overlay the produced portion of shale, and these layers serve as barriers to flow. In these circumstances, geologists and state regulators generally view as remote the possibility of creating a fracture that could reach a potable aquifer. If the shallow portions of shale formations were developed, then the thickness of the overlying rocks would be less and the distance from the shale to potable aquifers would be shorter, posing more of a risk to groundwater. In contrast to shale, coalbed methane (CBM) basins often qualify as underground sources of drinking water. Injection of fracturing fluids directly into or adjacent to such formations may be more likely to present a risk of contamination, and this is where initial regulatory attention and study was focused.[11]

State regulators have expressed more concern about the groundwater contamination risks associated with developing a natural gas or oil well (drilling through an overlying aquifer and casing, cementing, and completing the well), as opposed to hydraulic fracturing per se. The challenges of sealing off the groundwater and isolating it from possible contamination are common to the development of any oil or gas well, and are not unique to hydraulic fracturing. However, horizontally drilled, hydraulically fractured oil and gas wells pose more development and production challenges, and are subject to greater pressures than conventional vertical wells.

Identifying the source or cause of groundwater contamination can be difficult for various reasons, including the complexity of hydrogeologic processes and investigations, a lack of baseline testing of nearby water wells prior to drilling and fracturing, and the confidential business information status traditionally provided for fracturing compounds. Investigations by regulators and researchers generally have found that incidents involving residential water well contamination (including methane gas migration) have been caused by failure of well-bore casing and cementing or other well development and operating problems, rather than the hydraulic fracturing process.[12]

The debate over the groundwater contamination risks associated with hydraulic fracturing operations has been fueled in part by the lack of scientific studies to assess more thoroughly the current practices and related complaints and uncertainties. To help address this issue, Congress has asked the Environmental Protection Agency (EPA) to conduct a study on the relationship between hydraulic fracturing and drinking water.[13] The "hydraulic fracturing" debate also has been complicated by terminology. Many do not differentiate the well stimulation process of "fracing" or "fracking" from the full range of activities associated with unconventional oil and gas exploration and production.[14]

Other water quality concerns—associated with both conventional and unconventional oil and natural gas extraction—include the risks of contaminating ground and surface water from surface spills, leaks from pits, and siltation of streams from drilling and pad construction activities. Because of the large, but short-term, volumes of water needed for the hydraulic fracturing operations used to extract shale gas and tight oil, water consumption issues have emerged as well. Water use issues include the impacts that large water withdrawals might have on groundwater resources, streams and aquatic life (particularly during low-flow periods), and other competing uses (e.g., municipal or agricultural uses). Such impacts may be regional or localized, and can vary seasonally or with longer-term variations in precipitation.

The management of the large volumes of wastewater produced during natural gas production (including flowback from hydraulic fracturing operations and water produced from source formations) has emerged in many areas as a significant water quality issue, as well as a cost issue for producers. In some areas, such as portions of the Marcellus Shale region,[15] capacity is limited for wastewater disposal using underground injection wells (historically, the most common and preferred produced-water disposal practice in oil and natural gas fields), and surface discharge of wastewater is an increasingly restricted option.[16] Additionally, the injection of large volumes of wastewater into disposal wells has been associated with instances of induced seismicity.[17]

Air emissions associated with unconventional oil and natural gas production also have raised public health concerns and have drawn regulatory scrutiny. Air pollutants can be released during various stages of oil and natural gas production. Emission sources include pad, road, and pipeline construction; well drilling and completion, and flowback activities; and natural gas processing, storage, and transmission equipment. Key pollutants include methane (the main component of natural gas and a potent greenhouse gas), volatile organic compounds (VOCs), nitrogen oxides, sulfur dioxide, particulate matter, and various hazardous air pollutants.[18] According to EPA, the oil and gas industry is a significant source of methane and VOC emissions, which react with nitrogen oxides to form ozone (smog). EPA has identified hydraulically fractured gas wells during flowback as an additional source of these emissions in the natural gas industry.[19]

Releases of methane and other pollutants also can occur where natural gas is produced in association with oil, and natural gas gathering pipelines and other infrastructure are lacking. In such cases, the natural gas generally must be flared or vented. Flaring reduces VOC emissions compared to venting, but like venting, it contributes to greenhouse gas emissions without producing an economic value or displacing other fuel consumption.[20] Natural gas flaring has become an issue with the rapid and intense development of tight oil from the Eagle Ford Formation in Texas and the Bakken Formation in North Dakota, which have significant amounts of associated gas.[21] Other areas that have experienced large increases in tight oil production also have had increases in the amount of natural gas being flared.

State Regulation of Oil and Gas Development

Oil and natural gas development is occurring in at least 32 states.[22] Shale gas, tight oil, or other unconventional resources (such as coalbed methane) are

found in many of these states, primarily on non-federal lands (see Figure 6). States are the principal regulators of oil and gas production activities on state and private lands.[23] The federal government, through the Department of the Interior's Bureau of Land Management (BLM), has responsibility for overseeing oil and gas development on federally managed lands; however, some states require operators on federal public lands within state boundaries to comply with the state's oil and gas rules.[24]

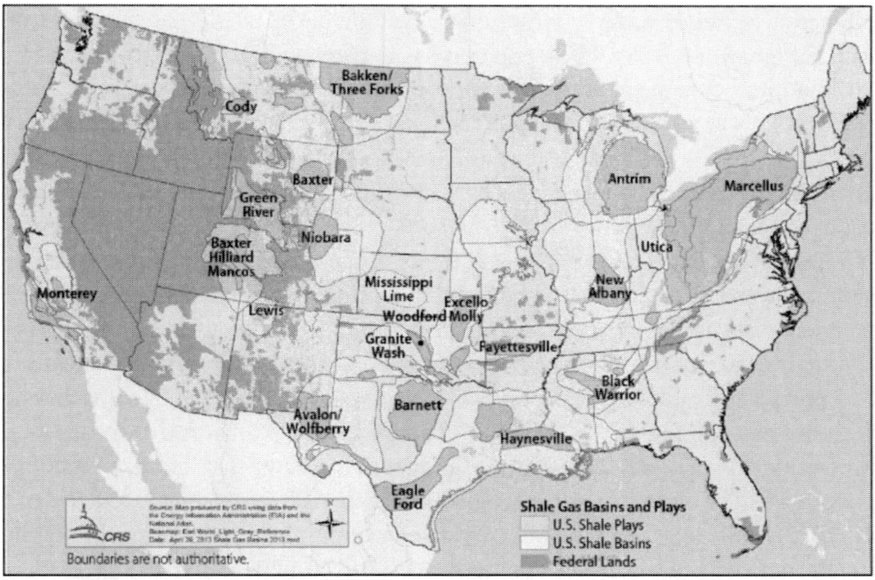

Source: CRS, compiled from U.S. Energy Information Administration sources.

Notes: No information had been reported on active shale plays in Alaska at the time of this report. Hawaii's volcanic origin does not support the geologic process leading to the deposition of shale.

Figure 6. Unconventional Shale Plays in the Lower 48 States, (with federal lands shown).

Hydraulic fracturing, traditionally without horizontal drilling, has been used for decades to stimulate increased production from existing oil or gas wells. This technique, along with other well stimulation techniques, has been regulated to varying degrees through state oil and gas codes. The detail and scope of applicable regulations vary across the states, and some states have regulated "well stimulation" broadly without addressing hydraulic "fracturing" explicitly.[25] State regulators have noted that hydraulic fracturing operations have been regulated through provisions that address various production activities, including requirements regarding well construction (e.g., casing and cementing), well stimulation (e.g., hydraulic fracturing), well operation (e.g., pressure testing and blowout prevention), and wastewater management.[26]

Nonetheless, drilling and fracturing methods and technologies have changed significantly over time as they have been applied to more challenging formations, greatly increasing the amount of water, fracturing fluids, and well pressures involved in oil and gas production operations. State groundwater protection officials have reported that development of shale gas and tight oil using high-volume hydraulic fracturing, in combination with directional drilling, has posed new challenges for the management and protection of water resources.[27] Consequently, many of the major producing states have revised or are in the process of revising their oil and gas laws and regulations to respond to these advances in oil and natural gas production technologies and related changes in the industry.[28]

When revising laws and regulations, states have added provisions to address hydraulic fracturing specifically, such as requirements for disclosure of chemicals used in hydraulic fracturing. Additionally, various states have adopted measures on water resources protection (including casing, cementing and pressure testing, well spacing, setbacks, water withdrawal, flowback, and wastewater storage and disposal requirements).[29] The Ground Water Protection Council reports that the number of states that have regulations governing hydraulic fracturing specifically increased from four in 2009 to 13 in 2013, and that the number of states requiring reporting of hydraulic fracturing chemicals grew from nine in 2009 to 21 in 2013.[30]

Taking a different approach, New York State has imposed a de facto moratorium on high-volume hydraulic fracturing pending completion of environmental and public health reviews and development of new rules. Similarly, Maryland regulators, pursuant to executive order, have studied the risks associated with deep drilling and hydraulic fracturing to identify new safeguards that may be needed in permits. In 2013, North Carolina lawmakers enacted legislation prohibiting the issuance of permits for oil and gas development using hydraulic fracturing and horizontal drilling until new regulations were in place and the legislature took affirmative action to allow permits to be issued, and in 2014, the state enacted legislation authorizing a regulatory permitting program for shale gas development.[31]

Debate over the Federal Role

While states continue to adopt and implement varying frameworks for oversight and regulation of unconventional gas and oil development, numerous citizen and environmental groups and Members of Congress have pressed for greater environmental oversight of shale energy development at the federal level. Some advocates of a larger federal role point to a wide range of differences in substance, scope, and enforcement among state regulatory regimes, and assert that a national framework is needed to ensure a consistent baseline level of environmental and human health protection and transparency.[32] Such advocates further argue that greater regulatory uniformity would reduce risks and uncertainties to both the industry and the public.[33] Others argue against greater federal involvement, and point to established state oil and gas programs and regulatory structures (which include a range of structures involving commissions, boards, or divisions within natural resource agencies working to varying degrees with, or within, state environmental agencies). In this view, experience lies with the states, and in addition to the relative nimbleness of states to review and revise laws and rules, the states are better able to consider regional differences in geology, topography, climate, and water resources.

In the 113th Congress, as in recent Congresses, the federal role in regulating oil and gas production generally, and hydraulic fracturing specifically, has been the subject of hearings, seminars, and legislation.[34] A number of bills have been proposed to broaden the federal role, while others have proposed to further limit federal involvement in regulating oil and gas development. Such proposals have been contentious, and Congress has not enacted such legislation since amending the Safe Drinking Water Act (SDWA) in the Energy Policy Act (EPAct) of 2005 (P.L. 109-58) to explicitly exclude from the SDWA definition of underground injection the injection of fluids (other than diesel fuels) related to hydraulic fracturing operations.[35]

Selected Federal Responses to Unconventional Resource Extraction

Provisions of several federal environmental laws and related regulations currently apply to certain activities associated with oil and natural gas production.[36] The Clean Water Act (CWA), for example, prohibits the discharge of pollutants from point sources into surface waters without a permit,[37] and the Safe Drinking Water Act (SDWA) requires an Underground Injection Control (UIC) permit for wastewater disposal through deep well injection.[38] Additionally, a SDWA UIC permit is required for the underground injection of fluids or propping agents pursuant to hydraulic fracturing if the injected fracturing fluids contain diesel fuels.[39] In 2012, EPA promulgated regulations under the authority of the Clean Air Act that require reductions in emissions related to oil and natural gas production, including emissions of volatile organic compounds (VOCs) from hydraulically fractured natural gas wells.[40]

While congressional debate has continued on legislative proposals, the Administration has been pursuing additional initiatives to regulate or otherwise manage activities related to unconventional oil and gas production. EPA has been most active, and is considering actions under several pollution control statutes. Among these efforts, EPA is working to (1) establish pretreatment standards to control discharges of wastewater from shale gas extraction to publicly owned wastewater treatment plants; (2) revise water quality criteria to protect aquatic life from discharges of brine produced during oil and gas extraction to surface waters; and (3) subject hydraulic fracturing chemicals to toxic substance reporting requirements.[41] In February 2014, EPA finalized permitting guidance for the use of diesel in hydraulic fracturing operations. The *Appendix* of this report provides a brief overview of selected federal environmental research and regulatory activities related to the production of tight oil and gas resources. Several of these initiatives are reviewed below.

EPA Study on Hydraulic Fracturing and Drinking Water

In 2009, the 111th Congress urged EPA to conduct a study on the relationship between hydraulic fracturing and drinking water to gain a better understanding of potential contamination risks.[42] In 2011, EPA published a final study plan that identified research projects that would address the full life cycle of water in hydraulic fracturing, from water acquisition to chemical mixing and injection through wastewater treatment and/or disposal. The study is intended to (1) examine conditions that may be associated with potential contamination of drinking water sources, and (2) identify factors that may lead to human exposure and risks.[43] As part of the study, EPA is investigating five reported incidents of drinking water contamination in areas where hydraulic fracturing has occurred. The purpose of the retrospective case studies is to determine the potential relationship between reported impacts and hydraulic fracturing activities.[44]

In December 2012, EPA released a status report presenting the agency's efforts through FY2012 on 18 research projects being conducted for the study.[45] No data or findings were included. EPA plans to synthesize the results from the research projects in a draft "report of results" in 2015. EPA has designated the report of results as a "highly influential scientific assessment" (HISA),[46] which will undergo peer review by EPA's independent Science Advisory Board.[47] In June 2013, an agency researcher stated that the final report will not be completed before 2016.

Multiagency Collaboration on Unconventional Oil and Gas Research

In March 2011, the White House issued a broad *Blueprint for a Secure Energy Future*, which identified a need to "expand safe and responsible domestic oil and gas development and production." Additionally, the President directed the Secretary of Energy to identify steps that could be taken to improve the safety and environmental performance of shale gas production, and to develop consensus recommendations on practices to ensure the protection of public health and the environment.[48]

In response, the Secretary of Energy's Advisory Board (SEAB) convened the Shale Gas Production Subcommittee to identify and evaluate issues and make recommendations to mitigate possible impacts of shale gas development. The final report included recommendations for the states, federal government, and industry. The subcommittee recommended, among other actions, that companies and regulators—to the extent that such actions had not been undertaken—adopt further measures to protect water quality and to manage water use and wastewater disposal, publicly report the composition of water and flow throughout the fracturing and cleanup process, disclose fracturing fluid composition, and adopt best practices for well development and construction (especially casing, cementing, and pressure management).[49] The committee also recommended actions to protect air quality through reduction of emissions of air toxics, ozone precursors, methane, and other pollutants.

In 2012, the President issued Executive Order (E.O.) 13605, *Supporting Safe and Responsible Development of Unconventional Domestic Natural Gas Resources*, to coordinate the efforts of federal agencies overseeing the development of unconventional domestic natural gas resources and associated infrastructure. The order states "Because efforts to promote safe, responsible, and efficient development of unconventional domestic natural gas resources are underway at a number of executive departments and agencies, close interagency coordination is important for effective implementation of these programs and activities."[50]

E.O. 13605 established an interagency working group to coordinate agency activities and to engage in long-term planning to ensure coordination on research, resource assessment, and infrastructure development. In April 2012, the lead agencies—the Department of Energy (DOE), EPA, and the Department of the Interior (DOI/U.S. Geological Survey)—signed a Memorandum of Agreement to develop a multiagency research plan "to address the highest priority research questions associated with safely and prudently developing unconventional shale gas and tight oil reserves." In July 2014, the three agencies released a research and development strategy for unconventional oil and gas resources.[51]

BLM Proposed Rule on Hydraulic Fracturing

While states have predominant regulatory authority for oil and gas development on state and private lands, the federal government is responsible for managing oil and gas resources on federal lands. However, some states require oil and gas operators on federal lands within their state to comply with state rules, and consequently, the debate over the federal role in regulating unconventional oil and gas production has extended to activities on federal lands.

The Bureau of Land Management (BLM), within the Department of the Interior, is the federal agency responsible for overseeing oil, natural gas, and coal leasing and production on federal and Indian lands, including split estate where the federal government owns the subsurface mineral estate.[52] BLM is tasked with leasing subsurface mineral rights not only on BLM-administered land, but also for lands managed by other federal agencies, including the U.S. Forest Service.[53] BLM oversees roughly 700 million subsurface acres of federal mineral estate and 56 million subsurface acres of Indian mineral estate nationwide. BLM estimates that approximately 3,400 wells have been drilled annually in recent years on federal and Indian lands, and that hydraulic fracturing is used to stimulate roughly 90% of these wells.[54]

In May 2012, BLM proposed revisions to its oil and natural gas development rules in response to the increased use of hydraulic fracturing on federal and Indian lands.[55] The proposed rule broadly addressed "well stimulation, including hydraulic fracturing," and would revise BLM oil and gas production regulations that were promulgated in 1982 and last revised in 1988.[56] In the 2012 *Federal Register* notice, BLM noted that the rule would modernize its management of well stimulation activities, and stated that the "rule is necessary to provide useful information to the public and to assure that hydraulic fracturing is conducted in a way that adequately protects the environment." [57] The preamble further noted that the proposed changes were partly in response to recommendations made by the aforementioned SEAB Shale Gas Subcommittee.

BLM received more than 177,000 comments on the proposed rule, and in May 2013, BLM published a Supplemental Notice of Proposed Rulemaking (SNPR) and Request for Comment. BLM has requested comments on the multiple changes in the proposed rule, and provided 30 days for public comment. (The comment period was extended for 60 days, to August 23, 2013.)[58] The bureau has responded to the roughly 1,340,000 comments it received on the SNPR, and has a goal of issuing a final rule in January 2015.

Changes notwithstanding, the 2012 proposed rule and the 2013 SNPR share overarching features that reflect recommendations of the SEAB subcommittee report. Both proposals would (1) add reporting and management requirements for water and other fluids used and produced in hydraulic fracturing operations, with emphasis on managing fluids that flow back to the surface, (2) require public disclosure of hydraulic fracturing chemicals, and (3) tighten well construction and operation requirements to help ensure that wellbore integrity is maintained throughout the hydraulic fracturing process.

Among the changes to the 2012 proposed rule, the BLM 2013 Supplemental Notice would

- narrow the scope of the rule to apply only to hydraulic fracturing and refracturing (the 2012 proposed rule would have applied to "well stimulation" activities broadly);[59]
- provide opportunities for individual states or tribes to work with BLM to craft variances for specific regulatory provisions that would allow compliance with state or tribal requirements to be accepted as compliance with the BLM rule (if the variance would meet or exceed the effectiveness of the rule provision it would replace);[60]
- allow operators to report hydraulic fracturing chemical information to BLM either directly or through the FracFocus website or other specified database,[61] and provide more detailed guidance on procedures for handling trade secret claims;[62]
- clarify that mechanical integrity testing would be required for all fracturing and refracturing operations;[63]
- require that all fracturing operations isolate all usable water formations to protect them from contamination, and allow operators to use an expanded set of cement evaluation tools to help ensure that usable water zones have been isolated and protected;[64] and

- allow an advanced Notice of Intent to be submitted for a single well, or group of wells with the same geological characteristics within a field where hydraulic fracturing operations are likely to be successful using the same design.[65]

BLM also requested comment on whether to require hydraulic fracturing wastewater to be stored in tanks only, rather than in lined pits or tanks as proposed in 2012. BLM sent the rule to the Office of Management and Budget (OMB) for review in 2014 and expects to promulgate a final rule in January 2015. The bureau also has begun taking steps to further revise its oil and gas rules to address emissions of air pollutants.

Coast Guard Regulation of Barge Shipments of Shale Gas Wastewater

The disposal of the large volumes of wastewater produced during shale gas extraction has posed challenges for companies, state regulators, and communities—particularly in the Marcellus Shale region. On-site disposal options are limited, and trucking wastewater to distant injection wells is costly. In 2012, the Coast Guard received two requests for approval for the bulk shipment of wastewater resulting from shale gas extraction in the Marcellus Shale to storage or treatment centers and final disposal sites in Ohio, Texas, and Louisiana.

The Coast Guard regulates the shipment of hazardous materials on the nation's rivers, and classifies cargoes for bulk shipment.[66] For a cargo that has not been classified in the regulations or under prior policy, the ship owner must request Coast Guard approval prior to shipping the cargo.[67] The Coast Guard has identified concerns with shipment of shale gas wastewater in barges. A key Coast Guard concern with the wastewater is "its potential for contamination with radioactive isotopes such as radium-226 and -228. Radium is of particular concern because it is chemically similar to calcium and so will easily form surface residues and may lead to radioactive surface contamination of the barges."[68] Consequently, the Coast Guard currently does not allow barge shipment of shale gas extraction wastewater (SGEWW), and is developing a policy to allow SGEWW to be transported for disposal. In March 2013, the Coast Guard submitted for review to OMB a draft document, "Carriage of Conditionally Permitted Shale Gas Extraction Waste Water in Bulk."

In October 2013, the Coast Guard published a notice of availability of a proposed "policy letter" concerning barge shipments of SGEWW and requested public comment. The Coast Guard received more than 70,000 comments, and has been reviewing them. After addressing public comments, the Coast Guard plans to issue a final policy letter that specifies conditions and information requirements that barge owners would be required to meet to receive approval to transport shale gas wastewater in bulk on inland waterways.[69]

LEGISLATION IN THE 113TH CONGRESS

Contrasting bills have been offered in the 113th Congress addressing unconventional oil and gas development, and hydraulic fracturing specifically. Several bills would expand federal regulation of hydraulic fracturing activities, while others would limit federal involvement.[70] House-passed H.R. 2728 would amend the Mineral Leasing Act[71] to prohibit the Department of the Interior from enforcing any federal regulation, guidance, or permit requirement regarding hydraulic fracturing relating to oil, gas, or geothermal production activities on or under any land in any state that has regulations, guidance, or permit requirements for hydraulic fracturing. Although this language is broadly applicable to any federal regulation, guidance, and permit requirements "regarding hydraulic fracturing," the prohibition on enforcement applies only to the Department of the Interior, and therefore would presumably impact only hydraulic fracturing operations on lands managed by that agency. The bill also would require the Department of the Interior to defer to state regulations, permitting, and guidance for all activities related to hydraulic fracturing relating to oil, gas, or geothermal production activities on federal land regardless of whether those rules were duplicative, more or less restrictive, or did not meet federal guidelines. The bill, as passed, would further prohibit the department from enforcing hydraulic fracturing regulations on Trust lands, except with express tribal consent. The House passed H.R. 2728, amended, on November 20, 2013. The same day, S. 1743, a companion bill to H.R. 2728, as introduced, was offered in the Senate. H.R. 2728 was placed on the Senate Legislative Calendar in December 2013. In September 2014, the House passed broad energy legislation (H.R. 2), which included the text of H.R. 2728 in Subdivision D.

Relatedly, the Fracturing Regulations are Effective in State Hands (FRESH) Act, H.R. 2513 and S. 1234, would establish that a state has sole authority to regulate hydraulic fracturing operations on lands within the boundaries of the state. The legislation further specifies that hydraulic fracturing on federal public lands shall be subject to the law of the state in which the land is located. H.R. 1548 (H.Rept. 113-263) would prohibit the BLM hydraulic fracturing rule from having any effect on land held in trust or restricted status for Indians, except with the express consent of its Indian beneficiaries. H.R. 2, Section 25009, includes this language. Similarly, S. 1482, the Empower States Act of 2013 generally would prohibit the Secretary of the Interior from issuing regulations or guidelines regarding oil and gas production on federal land in a state if the state has otherwise met the requirements under applicable federal law. Among other provisions, the bill also would (1) amend the Safe Drinking Water Act to require federal agencies, before issuing any oil and gas regulation or guideline, to seek comment and consult with each affected state agency and Indian tribe, and (2) require any future rule requiring disclosure of hydraulic fracturing chemicals to refer to the FracFocus database.

In contrast to the above bills, several others propose to expand federal regulation of hydraulic fracturing. In the first session, the Fracturing Responsibility and Awareness of Chemicals Act (FRAC) of 2013 was introduced in the House (H.R. 1921) and the Senate (S. 1135). The bills would amend the Safe Drinking Water Act to (1) require disclosure of the chemicals used in the fracturing process, and (2) repeal the hydraulic fracturing exemption established in EPAct 2005 and amend the term "underground injection" to include the injection of fluids used in hydraulic fracturing operations, thus authorizing EPA to regulate this process under the SDWA. The Climate Protection Act of 2013, S. 332, Section 301, contains similar chemical disclosure provisions. Additionally, S. 332 would repeal SDWA Section 1425, which provides states with an alternative to meeting the specific requirements contained in EPA UIC regulations promulgated under Section 1421 by allowing states to demonstrate to EPA that their existing programs for oil and gas injection wells are effective in preventing endangerment of underground sources of drinking water.[72] S. 332, Section 302, would require EPA to report to Congress on fugitive methane emissions resulting from natural gas infrastructure.

Legislation also has been introduced to require baseline and follow-up testing of potable groundwater supplies in the vicinity of hydraulic fracturing operations. H.R. 2983, the Safe Hydration is an American Right in Energy Development (SHARED) Act of 2013, would amend the SDWA to prohibit hydraulic fracturing unless the person proposing to conduct the fracturing operations agreed to testing and reporting requirements regarding underground sources of drinking water. The legislation would require testing prior to the start of injection operations, and during and after hydraulic fracturing operations. Testing would be required for any substance EPA determined would indicate damage associated with hydraulic fracturing operations. H.R. 2983 would require EPA to post on its website all test results, searchable by zip code.

H.R. 2850 (H.Rept. 113-252), the EPA Hydraulic Fracturing Study Improvement Act, would require EPA to follow certain procedures governing peer review and data presentation in conducting its study on the relationship between hydraulic fracturing and drinking water. As reported, the bill would require EPA to release the final report by September 30, 2016. The bill was included in Division D of H.R. 2, as passed by the House.

Broader oil and gas regulatory bills include H.R. 1154, the Bringing Reductions to Energy's Airborne Toxic Health Effects (BREATHE) Act, which would amend the Clean Air Act to authorize EPA to aggregate emissions from oil and gas wells, pipelines, and related units for purposes of regulating toxic air pollutants. H.R. 2825, the Closing Loopholes and Ending Arbitrary and Needless Evasion of Regulations (CLEANER) Act of 2013, would amend the Solid Waste Disposal Act to require EPA to determine whether wastes associated with oil and gas production meet the criteria for hazardous waste, and to regulate any such wastes.

Conclusion: Above- and Below-Ground Issues a Concern

The prospect that by the end of the decade the United States could become a significant exporter of natural gas and the world's leading oil producer is a phenomenal change of circumstances from just a few years ago. The technological advances that drove the changes in the United States have also reversed the global perspective of dwindling oil and natural gas resources, and increased the concern about greenhouse gas emissions. Other countries seek to emulate the U.S. production success, but have yet to do so. The U.S. oil and gas situation continues to be extremely dynamic, and many questions remain about how the United States will develop its resources.

Many observers, including U.S. government officials, have only recently recognized the tremendous resource size and the benefits that will accrue from developing the resources. Even though shale gas development is still considered very new and tight oil production is even newer, the industry has continued to improve its efficiency in extracting the resources, particularly of natural gas. As more industry resources are shifted to tight oil plays, the natural gas sector has had to produce more with less. Some in industry point out that at the beginning of shale gas development about 5% of the resource was able to be extracted; now it is closer to 20%, but will likely increase over time. By comparison, the extraction rate for conventional gas is between 30% and 60% of the resource.

Development of these resources has generated concern and debate over potential environmental and human health risks. Concerns include potential impacts to groundwater and surface water resources from well development and stimulation operations and wastewater management, as well as air quality impacts from emissions of air pollutants, including methane. These concerns have drawn scrutiny of regulatory regimes governing this industry, and have led to calls for greater federal oversight of oil and gas development. Although primary regulatory authority over oil and natural gas exploration and production on state and private lands generally rests with the states, provisions of several federal environmental laws currently apply to certain activities associated with oil and natural gas exploration and production. Moreover, EPA is reviewing other statutory authorities and pursuing new regulatory initiatives, and BLM has proposed revisions to its oil and gas rules to address hydraulic fracturing on federal and Indian lands. A broader concern among some is that the low price of natural gas is having negative consequences for the development and growth in energy efficiency, renewable energy sources, and nuclear power, potentially resulting in another generation of greenhouse-gas-producing energy sources.

The 113th Congress has held hearings, roundtables, and other discussions on issues associated with unconventional oil and gas development broadly, and on the role of the states specifically. Bills have been introduced to expand and also to constrain federal involvement in oil and gas development involving hydraulic fracturing. In the meantime, the Administration is pursuing actions to broaden federal oversight of this industry sector through administrative means.[73]

APPENDIX. SELECTED FEDERAL INITIATIVES RELATED TO UNCONVENTIONAL OIL AND GAS PRODUCTION

Table A-1. Selected Federal Actions Related to Unconventional Oil and Gas Production (with emphasis on hydraulic fracturing)

Agency: Statute, as Amended	Regulatory/Guidance	Research	Status
EPA: Clean Air Act (CAA)	Air emissions. In 2012, EPA issued regulations that revised existing rules and promulgated new ones to regulate emissions of volatile organic compounds (VOCs), sulfur dioxide, and hazardous air pollutants (HAPs) from many production and processing activities in the oil and gas sector that had not been subject previously to federal regulation.		Rules were promulgated in August 2012 (77 Federal Register 49489); requirements phase in through 2015.
	Particularly pertinent to shale gas production are the New Source Performance Standards (NSPS), which require reductions in emissions of VOCs from hydraulically fractured natural gas wells. The rules require operators to use reduced emissions completions (green completions) for all hydraulically fractured natural gas wells beginning no later than January 2015.		EPA agreed to revisit elements of the NSPS, and on April 12, 2013, proposed revisions to the NSPS for storage tanks (78 Federal Register 22125).
	Applying broadly across the sector, the NSPS require reductions of VOCs from compressors, pneumatic controllers, storage vessels, and other emission sources, and also revise existing standards for sulfur dioxide emissions from onshore natural gas processing plants, and HAPs from dehydrators and storage tanks.		

Table A-1. (Continued)

Agency: Statute, as Amended	Regulatory/Guidance	Research	Status
	In September 2013, EPA updated its 2012 performance standards for oil and natural gas to address VOC emissions from storage tanks used by the crude oil and natural gas production industry. The updates are intended to ensure tanks likely to have the highest emissions are controlled first, while providing tank owners and operators time to purchase and install VOC controls. The amendments reflect recent information showing that more storage tanks will be coming on line than the agency originally estimated (thus, presumably, producers need more time to purchase and install emission controls).[a]		On September 23, 2013, EPA finalized revisions to the NSPS for storage tanks (78 Federal Register 58416).
	In July 2014, EPA proposed updates and clarifications to NSPS requirements for well completions, storage tanks, and natural gas processing plants. The proposal would not change the required emission reductions in the rules, including standards applicable to hydraulically fractured natural gas wells.		On July 17, 2014, EPA proposed changes to the NSPS rules. (79 Federal Register 41752).
EPA: Clean Water Act (CWA)	Wastewater discharge. Produced water and flowback from hydraulic fracturing have high levels of total dissolved solids (TDS), largely chlorides, which can harm aquatic life and affect receiving water uses (such as fishing or irrigation). EPA is updating its chloride water quality criteria for protection of aquatic life.		Draft criteria document expected in late 2014.

Agency: Statute, as Amended	Regulatory/Guidance	Research	Status
	CWA Section 304(a)(1) requires EPA to develop criteria for water quality that reflect the latest scientific understanding of the effects of pollutants on aquatic life and human health. States use EPA-recommended criteria to establish state water quality standards, which in turn are used to develop enforceable discharge permits.		
	If reflected in state water quality standards, the revised chloride water quality criteria could affect discharges of produced water from extraction of conventional and unconventional oil and gas.[b]		
EPA: CWA	Wastewater discharge. In 2011, EPA indicated that it was initiating two separate rulemakings to revise the Effluent Limitations Guidelines and Standards (ELGs) for the Oil and Gas Extraction Point Source Category to control discharges of wastewater from (1) coalbed methane (CBM) and (2) shale gas extraction. Under CWA Section 304(m), EPA sets national standards for discharges of industrial wastewater based on best available technologies that are economically achievable (BAT). States incorporate these limits into discharge permits. Shale and CBM wastewaters often contain high levels of total dissolved solids (TDS—i.e., salts), and shale gas wastewater may contain chemical contaminants, naturally occurring radioactive materials (NORM), and metals.		Notice of the final Effluent Guidelines Program Plan was published in October 2011 (76 Federal Register 66286). For shale gas wastewater, EPA plans to propose a rule in February 2015, and finalize the rule in March 2016.

Table A-1. (Continued)

Agency: Statute, as Amended	Regulatory/Guidance	Research	Status
	Discharges to surface water: Currently, shale gas wastewater may not be discharged directly to surface waters. CBM wastewater is not subject to national discharge standards; rather, CBM wastewater discharge permits are based on best professional judgments of state or EPA permit writers. EPA was working to develop regulatory options to control direct discharges of CBM wastewaters, but determined in 2013 that no economically achievable technology was available. *Discharges to treatment plants:* Current ELGs lack pretreatment standards for discharges of shale gas or CBM wastewaters to publicly owned wastewater treatment works (POTWs), which typically are not designed to treat this wastewater. EPA is developing national pretreatment standards that shale gas and CBM wastewaters would be required to meet before discharge to a POTW to ensure that the receiving facility could treat the wastewater effectively.[c]		On August 7, 2013, EPA proposed to delist CBM from the ELG rulemaking plan based on the "declining prevalence and economic viability" of the industry. EPA determined that no economically achievable technology is available currently (78 Federal Register 48159).
EPA: Emergency Planning and Community Right-to-Know Act (EPCRA)	Chemical disclosure. EPA has been considering an October 2012 petition by nongovernmental organizations to subject the oil and natural gas extraction industry to Toxics Release Inventory (TRI) reporting under EPCRA. Section 313 of EPCRA requires owners or operators of		Notice of receipt of petition published on January 3, 2014 (79 Federal Register 393). No published schedule for EPA's response

Agency: Statute, as Amended	Regulatory/Guidance	Research	Status
	certain industrial facilities to report on releases of toxic substances to the state and EPA. EPA and states are required to make nonproprietary data publicly available through the TRI website.		to petition.
EPA: Safe Drinking Water Act (SDWA)	Diesel fuels. EPA has issued *UIC Program Guidance for Permitting Hydraulic Fracturing with Diesel Fuels* in response to the revised SDWA definition of "underground injection" in the Energy Policy Act (EPAct) of 2005 to explicitly exclude the underground injection of fluids (other than diesel fuels) used in hydraulic fracturing. The guidance provides recommendations for EPA permit writers to use in writing permits for hydraulic fracturing operations using diesel fuels. The guidance applies in states where EPA implements the UIC program for oil and natural gas related (Class II) injection wells. States are not required to adopt the guidance, but may do so.[d]		Draft guidance issued in May 2012. Final guidance issued in February 2014.
EPA: SDWA		Study. EPA is studying the relationship between hydraulic fracturing and drinking water. Congress requested the study in EPA's FY2010 appropriations act. EPA designated the	Progress report issued in December 2012. Draft report is expected to be submitted for peer review in 2015. A final report is expected in 2016 (extended from 2014).

Table A-1. (Continued)

Agency: Statute, as Amended	Regulatory/Guidance	Research	Status
		pending "report of results" as a "highly influential scientific assessment" (HISA), which requires peer review by qualified specialists.	
EPA: Toxic Substances Control Act (TSCA)	Chemical reporting. In response to a citizen petition (TSCA Section 21), EPA published an Advance Notice of Proposed Rulemaking (ANPRM) to get input on the design and scope of possible reporting requirements for hydraulic fracturing chemicals. EPA is considering requiring information reporting under TSCA Section 8(a), and health and safety data reporting under Section 8(d). EPA is seeking comment on the types of chemical information that could be reported and disclosed, and approaches to obtaining this information for chemicals used in hydraulic fracturing.		Initiated in January 2012. Advanced Notice of Proposed Rulemaking (ANPR) under TSCA Section 8 published May 9, 2014 (79 Federal Register 28664). Public comment period closed September 18, 2014.
EPA: Resource Conservation and Recovery Act (RCRA)	Storage/disposal pits and ponds. EPA has been considering developing guidance to address the design, operation, maintenance, and closure of pits used to store hydraulic fracturing fluids for reuse or pending final disposal. These wastes are exempt from regulation as a hazardous waste under RCRA. In April 2014, EPA issued a		In April 2014, EPA issued a Compilation of Publicly Available Sources of Voluntary Management Practices for Oil and Gas Exploration and

Agency: Statute, as Amended	Regulatory/Guidance	Research	Status
	document that compiles voluntary management practices for oil and gas exploration and production wastes. This nonregulatory, non-guidance document is intended to provide information only, and does not establish agency policy.		Production (E&P) Wastes as They Address Pits, Tanks, and Land Application.
Department of the Interior, Bureau of Land Management (BLM): Mineral Leasing Act, Indian Mineral Leasing Act	Hydraulic fracturing on public lands. BLM has proposed revisions to rules governing oil and natural gas production on federal and Indian lands. BLM proposes to (1) require public disclosure of chemicals used in hydraulic fracturing, (2) tighten regulations related to well-bore integrity, and (3) add new reporting and management/storage/disposal requirements for water used in hydraulic fracturing.		Rule was first proposed in May 2012; after extensivepublic comment, BLM issued a Supplemental Notice of Proposed Rulemaking on May 24, 2013 (78 Federal Register 31636). Final rule expected in January 2015.
Department of Homeland Security, Coast Guard: 46 U.S.C. Ch. 37	Wastewater shipment. The Coast Guard regulates the shipment of hazardous materials on the nation's rivers. Because of the potential for shale gas wastewater in the Marcellus Shale region to contain radioactive materials (especially radium, which can form surface residues and may lead to radioactive surface contamination of the barges), the Coast Guard currently does not allow barge shipment of shale gas extraction wastewater. In 2013, the Coast Guard's Hazardous Materials Division issued a proposed policy letter establishing requirements for bulk shipment of shale gas extraction wastewater by barge for disposal.		On October 30, 2013, the Coast Guard published a notice for a onemonth comment period on a proposed policy letter setting conditions for bulk shipment of shale gas wastewater (78 Federal Register 64905).

Table A-1. (Continued)

Agency: Statute, as Amended	Regulatory/Guidance	Research	Status
	The Coast Guard received more than 70,000 comments, and has been reviewing them.		
DOE/EPA/DOI-USGS: E.O. 13605		Federal research coordination. In 2012, the three agencies agreed, through an MOU, to develop a multiagency research plan "to address the highest priority research questions associated with safely and prudently developing unconventional shale gas and tight oil resources."	Multiagency Research Strategy was issued on July 18, 2013.[e]

Source: Prepared by the Congressional Research Service.

Notes: This table presents selected Administration activities related to unconventional oil and natural gas extraction. It excludes, for example, regional or site-specific research studies conducted by federal agencies. More information on EPA initiatives to regulate oil and gas production and hydraulic fracturing is available at EPA's website, Natural Gas Extraction—Hydraulic Fracturing, http://www2.epa.gov/hydraulicfracturing.

[a] <001These CAA rules, issued under court order, establish new air emissions standards for the "Crude Oil and Natural Gas Production" and "Natural Gas Transmission and Storage" source categories. For details, see CRS Report R42986, An Overview of Air Quality Issues in Natural Gas Systems, by Richard K. Lattanzio.

[b] <003For more information, see the EPA Water Quality Criteria web page, http://water swguidance/standards/criteria/.

[c] <004EPA explains that "[f]or direct dischargers of unconventional oil and gas wastewaters from onshore oil and gas facilities—with the exception of coalbed

methane—technology-based limitations are based on the Effluent Limitations Guidelines (ELGs) for the Oil and Gas Extraction Category (40 CFR Part 435). Permits for onshore oil and gas facilities must include the requirements in Part 435, including a ban on the discharge of pollutants, except for wastewater that is of good enough quality for use in agricultural and wildlife propagation for those onshore facilities located in the continental United States and west of the 98th meridian.... Part 435 does not currently include categorical pretreatment standards for indirect discharges to publicly owned treatment works (POTWs) for wells located onshore." Source: U.S. Environmental Protection Agency, Unconventional Extraction in the Oil and Gas Industry, http://water.epa.gov/ scitech/ wastetech/guide/oilandgas/unconv.cfm.

[d] <006EPA regulates the underground injection of fluids through SDWA §§1421-1426; 42 U.S.C. §§300h-300h-5. In February 2014, EPA issued UIC Program Guidance for Permitting Hydraulic Fracturing with Diesel Fuels, which generally follows EPA Class II underground injection well requirements (i.e., well construction standards; mechanical integrity testing; operating, monitoring, and reporting requirements; and public notification and financial responsibility requirements). The guidance provides recommendations for EPA permit writers for tailoring requirements for hydraulic fracturing using diesel fuels. The guidance applies in states where EPA implements the UIC program for Class II wells (including Pennsylvania, New York, Michigan, Kentucky, Tennessee, and Virginia).

[e] Federal Multiagency Collaboration on Unconventional Oil and Gas Research—A Strategy for Research and Development, http://unconventional.energy

End Notes

[1] Hydraulic fracturing is an industry technique that uses water, sand, and chemicals under pressure to enhance the recovering of natural gas and oil. It has taken on new prominence as it has been applied to tight oil and shale gas formation as an essential method for producing resources from those types of formations.

[2] Coalbed methane and tight gas accounted for 33% of U.S. natural gas production in 2011, but are projected to account for only 28% in 2040, according to the Energy Information Administration (EIA).

[3] Full Committee Forum: Domestic Supply and Exports, Senate Energy and Natural Resources Committee, May 21, 2013, http://www.energy.senate.gov/public/index.cfm/hearings-and-business-meetings?ID=0380bed7-f9ef-4450-bfa0- a3af60f7a184.

[4] "U.S. Energy Abundance: Regulatory, Market, and Legal Barriers to Export," House Energy and Commerce Committee, Subcommittee on Energy and Power, June 18, 2013, http://energycommerce.house.gov/hearing/us-energyabundance-regulatory-market-and-legal-barriers-export.

[5] For additional information on U.S. natural gas exports, see CRS Report R42074, U.S. Natural Gas Exports: New Opportunities, Uncertain Outcomes, by Michael Ratner et al.

[6] For additional information on the Bakken Formation, see CRS Report R42032, The Bakken Formation: Leading Unconventional Oil Development, by Michael Ratner et al.

[7] BP, BP Statistical Review of World Energy, June 2014, pp. 8-9.

[8] CRS Report R41875, The U.S.-Canada Energy Relationship: Joined at the Well, by Paul W. Parfomak and Michael Ratner. Mexico is the third-largest source of U.S. oil imports, but is not always included in discussions of North American energy independence, as its oil sector is not as integrated with the United States as is Canada's.

[9] U.S. Energy Information Administration, U.S. Imports of Crude Oil and Petroleum Products, July 30, 2014, http://www.eia.gov/dnav/pet/hist/LeafHandler.ashx?n=PET&s=MCRIMUS2&f=A.

[10] Adam Sieminski, Outlook for Shale Gas and Tight Oil Development in the U.S., U.S. Energy Information Administration, Presentation for the American Petroleum Institute, Washington, DC, April 4, 2013, p. 12.

[11] U.S. Environmental Protection Agency, Evaluation of Impacts to Underground Sources of Drinking Water by Hydraulic Fracturing of Coalbed Methane Reservoirs, Final Report, EPA-816-04-003, Washington, DC, June 2004, p. 4-1. (EPA reviewed 11 major coalbed methane formations to determine whether coal seams lay within underground sources of drinking water (USDWs). EPA determined that 10 of the 11 producing coal basins "definitely or likely lie entirely or partially within USDWs.")

[12] Avner Vengosh, Robert B. Jackson, and N. Warner, et al., "A Critical Review of the Risks to Water Resources from Unconventional Shale Gas Development and Hydraulic Fracturing in the United States," Environmental Science and Technology, vol. 10, no. 1021 (2014), p. 405118.

[13] Department of the Interior, Environment, and Related Agencies Appropriations Act, 2010, P.L. 111-88, H.Rept. 111- 316. The EPA study (expected to be published in 2016) includes five case studies that involve drinking water contamination incidents in areas where unconventional oil and gas development is occurring.

[14] A 2012 Pacific Institute study found that many individuals interviewed for the study defined "hydraulic fracturing" much more broadly than the industry meaning of the term (i.e., the pressurized injection of fluids into a production well). These individuals used the term broadly to include well construction, completion, and other associated activities. Noting the differences, the authors concluded that "additional work is needed to clarify terms and definitions associated with hydraulic fracturing to support more fruitful and informed dialog and to develop appropriate energy, water, and environmental policy." See Hydraulic Fracturing and Water Resources: Separating the Frack from the Fiction, p. 29, http://www.pacinst.org/reports/fracking/.

[15] The Marcellus Shale formation is one of the largest unconventional natural gas resources in the United States, underlying much of West Virginia and Pennsylvania, southern New York, eastern Ohio, western Maryland, and western Virginia.

[16] For a discussion of water management issues associated with shale energy development, see CRS Report R43635, Shale Energy Technology Assessment: Current and Emerging Water Practices, by Mary Tiemann, Peter Folger, and Nicole T. Carter. See also Jeffrey Logan, Garvin Heath, and Jordan Macknick, et al., Natural Gas and the Transformation of the U.S. Energy Sector: Electricity, National Renewable Energy Laboratory, NREL Report No. TP-6A50-55538, November 2012, 225 pp., http://www.nrel.gov/docs/fy13osti/55538.pdf.

[17] William L. Ellsworth, "Injection-Induced Earthquakes," Science, vol. 341, July 12, 2013, http://www.sciencemag.org/content/341/6142/1225942.full.

[18] For a detailed discussion of air pollution issues associated with oil and gas exploration and development and recent EPA regulations, see CRS Report R42833, Air Quality Issues in Natural Gas Systems, by Richard K. Lattanzio.

[19] U.S. Environmental Protection Agency, Overview of Final Amendments to Air Regulations for the Oil and Natural Gas Industry: Fact Sheet, EPA, October 2012, http://www.epa.gov/airquality/oilandgas/pdfs/20120417fs.pdf.

[20] When vented, natural gas (largely methane) is released to the air without being burned. In contrast, when natural gas is flared (burned), the main by-product is carbon dioxide. Flaring is preferred to venting for safety reasons, but also because methane is several times more potent than carbon dioxide as a greenhouse gas (although more short-lived in the atmosphere). Flaring also reduces emissions of ozone-forming pollutants, compared to venting.

[21] See CRS Report R42032, The Bakken Formation: Leading Unconventional Oil Development, by Michael Ratner et al. See also U.S. Energy Information Administration, "North Dakota Aims to Reduce Natural Gas Flaring," Today in Energy, October 20, 2014, http://www.eia.gov/todayinenergy/detail.cfm?id=18451.

[22] U.S. Energy Information Administration, Rankings: Natural Gas Marketed Production, 2012, http://www.eia.gov/ state/rankings/#/series/47. EIA reports gas production in 32 states and oil production in 31 states. Five states (Texas, North Dakota, California, Alaska, and Oklahoma) accounted for the bulk of oil and gas production in 2012. The biggest gains in oil production were in North Dakota and Texas, due in large part to increased horizontal drilling and hydraulic fracturing activity.

[23] For a review of federal laws and regulations addressing leasing of federal lands for exploration and production of oil, gas, and coal, see CRS Report R40806, Energy Projects on Federal Lands: Leasing and Authorization, by Adam Vann.

[24] States often enter into memoranda of understanding with BLM to coordinate administration and enforcement of various regulatory requirements on public lands within the state.

[25] For state-specific information, see the Interstate Oil and Gas Compact Commission, Summary of State Statutes and Regulations, available at http://www.iogcc.state.ok.us/state-statutes.

[26] For example, before the state enacted hydraulic fracturing legislation (SB 4) in September 2013, California regulators noted that requirements for protecting underground resources and well construction standards "provide a first line of protection from potential damage caused by hydraulic fracturing." However, the state noted "There is a gap between the requirements placed on oil and gas operators to safely construct and maintain their wells, and the information they provide to the Division about hydraulic fracturing operations and steps taken to protect resources and the environment. The Department's pending regulatory process is intended to close that gap." California Department of Conservation, Hydraulic Fracturing in California, http://www.conservation.ca.gov/dog/general_information/Pages/HydraulicFracturing.aspx. Among other provisions, the California law requires public disclosure of chemicals, baseline and follow-up testing of nearby water wells, notification to nearby property owners and tenants, and groundwater monitoring plans, and directs the state to conduct a comprehensive environmental study of impacts associated with hydraulic fracturing. Also, SB 4 directs regulators to make any needed revisions to rules governing construction of wells and well casings to ensure well integrity.

[27] See, for example, Ground Water Protection Council, State Oil and Natural Gas Regulations Designed to Protect Water Resources, 2014, http://www.gwpc.org/state-oil-gas-regulations-designed-protect-water-resources-2014-edition.

[28] Alabama, Alaska, Arkansas, California, Colorado, Michigan, Montana, North Dakota, New Mexico, Ohio, Pennsylvania, Texas, Utah, West Virginia, and Wyoming are among the states that in recent years have revised oil and gas laws and/or rules that address unconventional oil and gas development, and hydraulic fracturing specifically. Among states currently revising oil and gas rules are California, Indiana, Maryland, New York, and North Carolina.

[29] For a comparison of state requirements for specific activities (e.g., wastewater disposal, chemical disclosure, and cementing), see Resources for the Future, A Review of Shale Gas Regulations by State, Center for Energy Economics and Policy, July 2012, http://www.rff.org/centers/energy_economics_and_policy/Pages/Shale_Maps.aspx.

[30] Ground Water Protection Council, State Oil and Natural Gas Regulations Designed to Protect Water Resources, 2014, p. 8.

[31] General Assembly of North Carolina, Session Law 2013-365, Senate Bill 76. Also, Vermont banned hydraulic fracturing (DOE's EIA does not list Vermont as an oil- or gas-producing state). The New Jersey legislature passed a ban on shale gas drilling; however, the governor vetoed the bill and imposed a one-year ban, which has expired.

[32] See, for example, Matthew McFeeley, State Hydraulic Fracturing Disclosure Rules and Enforcement: A Comparison, Natural Resources Defense Council, NRDC Issue Brief IB: 12-06-A, July 2012.

[33] For further discussion, see Jeffrey Logan, Garvin Heath, and Elizabeth Paranhos et al., Natural Gas and the Transformation of the U.S. Energy Sector: Electricity, Joint Institute for Strategic Energy Analysis, NREL/TP-6A50- 57702, Golden, CO, January 2013, http://www.nrel.gov/docs/fy13osti/55538.pdf.

[34] The 113th Congress has explored the role of states and the federal government in oil and gas production, specifically, and in environmental protection broadly. In February 2013, the House Committee on Energy and Commerce, Subcommittee on Environment and the Economy, held a hearing, The Role of the States in Protecting the Environment Under Current Law. The Senate Committee on Energy and Natural Resources held a series of Natural Gas Roundtables, including a May 2013 forum on Shale Development: Best Practices and Environmental Concerns.

[35] The Safe Drinking Water Act requires regulation of underground injection activities to protect underground sources of drinking water. EPA has long regulated underground injections related to oil and gas field wastewater disposal and enhanced oil recovery. Historically, EPA had not regulated injection of fluids for hydraulic fracturing of oil or gas production wells. In 1997, the U.S. Court of Appeals for the 11th Circuit ruled that injections for fracturing for coalbed methane production in Alabama constituted underground injection and must be regulated under the SDWA. For more information, see CRS Report R41760, Hydraulic Fracturing and Safe Drinking Water Act Regulatory Issues, by Mary Tiemann and Adam Vann.

[36] See CRS Report R43152, Hydraulic Fracturing: Selected Legal Issues, by Adam Vann, Brandon J. Murrill, and Mary Tiemann.

[37] CWA Section 301 prohibits the discharge of pollutants into the nation's waters except in compliance with the provisions of the law, which include obtaining a discharge permit. 33 U.S.C. §1311. For information on applicable CWA requirements, see Environmental Protection Agency, "Natural Gas Drilling in the Marcellus Shale, NPDES Program Frequently Asked Questions," March 16, 2011, http://www.epa.gov/npdes/pubs/hydrofracturing_faq.pdf.

[38] The Safe Drinking Water Act of 1974 (P.L. 93-523) authorized the Underground Injection Control (UIC) program at EPA. UIC provisions, as amended, are contained in SDWA Part C, §§1421-1426; 42 U.S.C. §§300h-300h-5.

[39] EPAct 2005 (P.L. 109-58, §322), amended SDWA to exempt from the definition of underground injection the injection of fluids or propping agents (other than diesel fuel) for hydraulic fracturing purposes.

[40] The rules regulate VOC emissions from hydraulically fractured natural gas wells, compressors, pneumatic controllers, storage vessels, and leaking components at onshore natural gas processing plants, as well as sulfur dioxide emissions from onshore natural gas processing plants. The new standards require producers to capture about 90% of the natural gas that escapes into the atmosphere as a result of production using hydraulic fracturing. For further discussion, see CRS Report R42986, An Overview of Air Quality Issues in Natural Gas Systems, by Richard K. Lattanzio.

[41] For details, see CRS Report R43152, Hydraulic Fracturing: Selected Legal Issues, by Adam Vann, Brandon J. Murrill, and Mary Tiemann.

[42] The Department of the Interior, Environment, and Related Agencies Appropriations Act, 2010 (P.L. 111-88, H.Rept. 111-316):

Hydraulic Fracturing Study.—The conferees urge the Agency to carry out a study on the relationship between hydraulic fracturing and drinking water, using a credible approach that relies on the best available science, as well as independent sources of information. The conferees expect the study to be conducted through a transparent, peer-reviewed process that will ensure the validity and accuracy of the data. The Agency shall consult with other Federal agencies as well as appropriate State and interstate regulatory agencies in carrying out the study, which should be prepared in accordance with the Agency's quality assurance principles.

[43] U.S. Environmental Protection Agency, Plan to Study the Potential Impacts of Hydraulic Fracturing on Drinking Water Sources, Office of Research and Development, EPA/600/R-11/122, November 2011, http://www2.epa.gov/ hfstudy.

[44] EPA is conducting retrospective case studies at five sites to develop information about the potential impacts of hydraulic fracturing on drinking water resources under different circumstances. The case studies include (1) the Bakken Shale in Dunn County, ND; (2) the Barnett Shale in Wise County, TX; (3) the Marcellus Shale in Bradford County, PA; (4) the Marcellus Shale in Washington County, PA; and (5) coalbed methane in the Raton Basin, CO.

[45] U.S. Environmental Protection Agency, Study of the Potential Impacts of Hydraulic Fracturing on Drinking Water Resources: Progress Report, Office of Research and Development, EPA 601/R-12/011, December 2012.

[46] Ibid., p. 4.

[47] Because EPA has designated the "report of results" as a "highly influential scientific assessment," the agency is to follow the peer review planning requirements described in the Office of Management and Budget's Information Quality Bulletin for Peer Review, 2004. The Bulletin states that important scientific information must be peer reviewed by qualified specialists before being disseminated by the federal government. The EPA Science Advisory Board is an external federal advisory committee that conducts peer reviews of significant EPA research products and activities.

[48] The White House, "Blueprint for a Secure Energy Future," March 30, 2011, p. 13, http://www.whitehouse.gov/sites/ default/files/blueprint_secure_energy_future.pdf.

[49] U.S. Department of Energy, the Secretary of Energy Advisory Board (SEAB), Shale Gas Production Subcommittee, Second Ninety Day Report, November 18, 2011, http://www.shalegas.energy.gov/. In November 2013, Secretary Moniz requested the SEAB to form a task force to review how FracFocus "houses the information Federal and State regulatory agencies require as part of their regulatory functions with regard to disclosure of the composition and quantities of fracturing fluids injected into unconventional oil and gas wells." This review is available at http://energy.gov/seab/ secretary-energy-advisory-board-seab-task-force-fracfocus-20.

[50] Executive Order 13605, "Supporting Safe and Responsible Development of Unconventional Domestic Natural Gas Resources" (Washington: GPO, 2012), http://www.gpo.gov/ fdsys/pkg/DCPD-201200269/pdf/DCPD-201200269.pdf.

[51] The Memorandum of Agreement and research strategy are available at the Administration website, "Multi-Agency Collaboration on Unconventional Oil and Gas Research," http://unconventional.energy.gov/.

[52] Mineral Leasing Act of 1920, 30 U.S.C. §181.

[53] For a discussion of federal lands leasing authorities and activities, see CRS Report R40806, Energy Projects on Federal Lands: Leasing and Authorization, by Adam Vann.

[54] 77 Federal Register 27691-27693.

[55] Department of the Interior, Bureau of Land Management, "Oil and Gas; Well Stimulation, Including Hydraulic Fracturing, on Federal and Indian Lands," 77 Federal Register 27691, May 11, 2012.

[56] The proposed rule would amend existing BLM regulations at 43 C.F.R. Section 316.3-2.

[57] 77 Federal Register 27692-27693.

[58] Department of the Interior, Bureau of Land Management, "Oil and Gas; Hydraulic Fracturing on Federal and Indian Lands: Supplemental Notice of Proposed Rulemaking," 78 Federal Register 31636, May 24, 2013.

[59] SNPR, §3162.3-3(a) and §3160.0-5. This revision excludes acidizing and enhanced secondary and tertiary recovery so that the rule would apply only to hydraulic fracturing and not to other "well stimulation" activities.

[60] SNPR, §3162.3-3(k). In 2012, BLM proposed to implement on public lands "whichever rules, state or Federal, are most protective of Federal lands and resources and the environment, consistent with longstanding practice and relevant statutory authorities." 77 Federal Register 72694.

[61] SNPR, §3162.3-3(i). Operators submitting information through FracFocus would be required to certify that the information is correct and certify that the operator complied with applicable laws governing notice and permits. FracFocus was established in 2011 by the Ground Water Protection Council and the Interstate Oil and Gas Compact Commission (IOGCC). FracFocus is a public registry where oil and gas companies may voluntarily identify chemicals used in hydraulic fracturing operations at specific wells. Many states allow or require operators to meet state disclosure requirements by posting information on the FracFocus website (http://www.fracfocus.org).

[62] SNPR, §3162.3-3(j)1-4. Modeled on Colorado rules, BLM would instruct operators not to disclose trade secret information to BLM or on FracFocus. Operators would submit an affidavit stating that withheld information is entitled to withholding. BLM would retain authority to require operators to submit claimed trade secret information.

[63] SNPR, §3162.3-3(f).

[64] SNPR, §3162.3-3(b) and §3162.3-3(d).

[65] SNPR, §3162.3-3(d).

[66] This action is based on authority in 46 U.S.C. Chapter 37—"Carriage of Liquid Bulk Dangerous Cargoes." Implementing regulations are published in 46 C.F.R. Subchapter O—"Certain Bulk Dangerous Cargoes."

[67] See 46 C.F.R. 153.900(c)-(d) or 46 C.F.R. 151.01-15.

[68] U.S. Coast Guard, Marine Safety Engineering, Shale Gas Extraction Waste Water, Commercial Regulations and Standards Directorate, Fall 2012, p. 5, http://www.uscg. mil/hq/cg5/cg52/docs/2012fall.pdf.

[69] Department of Homeland Security, Coast Guard, "Carriage of Conditionally Permitted Shale Gas Extraction Waste Water in Bulk: Notice of Availability and Request for Comments," 78 Federal Register 64905, October 30, 2013. http://www.uscg.mil/hq/cg5/cg521/.

[70] Similar bills were offered in the 112th Congress, but none was enacted. H.R. 1084/S. 587 proposed repealing the hydraulic fracturing exemption established in the Energy Policy Act (EPAct) of 2005, and amending the term "underground injection" to include the injection of fluids used in hydraulic fracturing operations, thus authorizing EPA to regulate this process under the SDWA. The bills also would have required disclosure of the chemicals used in the fracturing process. In response to rules proposed by BLM in 2012, S. 2248/H.R. 4322 proposed that a state would have sole authority to regulate hydraulic fracturing on federal lands within state boundaries; H.R. 3973 would have prohibited the rule from having any effect on Indian lands; and H.R. 6235 would have barred a final rule for 10 years, pending an impact study. For further discussion, see CRS Report R41760, Hydraulic Fracturing and Safe Drinking Water Act Regulatory Issues, by Mary Tiemann and Adam Vann.

[71] 30 U.S.C. §181 et seq.

[72] 42 U.S.C. §300h-4.

[73] See the Appendix for a review of federal research and regulatory initiatives related to unconventional oil and gas production, with emphasis on hydraulic fracturing.

In: Unconventional Oil and Shale Gas
Editor: Amber L. Tuft
ISBN: 978-1-63482-130-8

Chapter 2

SHALE ENERGY TECHNOLOGY ASSESSMENT: CURRENT AND EMERGING WATER PRACTICES*

Mary Tiemann, Peter Folger and Nicole T. Carter

SUMMARY

Shale oil and gas (collectively referred to as shale energy), long considered "unconventional" hydrocarbon resources, are now being developed rapidly. Economic extraction of shale energy resources typically relies on the use of hydraulic fracturing. This technique often requires significant amounts of freshwater, and fracturing flowback and related wastewaters must be recycled or disposed of after a well is completed. While shale energy presents a significant energy resource, its development has the potential to pose risks to water availability and water quality.

This report provides a technological assessment of existing and emerging water procurement and management practices in shale energy-producing regions of the United States. The intersection of evolving technology, growing environmental concerns, demand for new sources of hydrocarbon energy, and the potential national interests in developing shale oil and gas resources provides the context for this study. Congressional attention has been focused on two key aspects of the issue:

---

* This is an edited, reformatted and augmented version of a Congressional Research Service publication, No. R43635, dated July 14, 2014.

shale energy as a growing U.S. energy source, and environmental concerns associated with the development of these resources.

Water for shale energy projects is used most intensely in the fracturing portion of a well's life cycle. Under current practices, fracturing typically is a water-dependent activity, often requiring between a few million and 10 million gallons of water per fractured horizontal well. This water demand often is concentrated geographically and temporally during the development of a particular shale formation. Production activities and management and treatment of the wastewater produced during shale energy production (including flowback from fracturing and water produced from source formations) have raised concerns over the potential contamination of groundwater and surface water and induced seismicity associated with wastewater injection wells.

Water resource issues may pose constraints on the future development of domestic shale oil and gas. Potential negative effects from shale energy extraction—particularly effects associated with hydraulic fracturing and wastewater management—have prompted state and regional regulatory actions to protect water supplies. Future congressional and executive branch actions may influence development of shale oil and shale gas on federal lands and elsewhere through additional regulatory oversight or other policy actions. At the same time, advances in shale energy extraction and wastewater management techniques may reduce some development impacts.

The pace of technological change in water sourcing and water management in the shale energy sector is rapid, but uneven. Trends in water management have generally been influenced by local disposal costs, regulations, and geologic conditions rather than by water scarcity alone. Emerging technologies and practices in water resources management can be divided into those that seek to reduce the amount of consumptive freshwater utilization in the drilling and completion process, and those that seek to lower the costs and/or minimize the potential for negative environmental impacts associated with wastewater management.

Water management issues are relevant to the entire life cycle of shale energy development, because fluids will continue to be produced even after a well is drilled, fractured, and producing oil and/or natural gas. Research that views the shale energy production process in a life-cycle and materials-flow context may facilitate the identification of technologies and processes that can mitigate potential impacts along different stages of shale energy development.

INTRODUCTION

This report provides an assessment of current and emerging water procurement and management technologies and practices related to shale energy development in the United States. Water resource management issues associated with shale energy development are of concern to policy makers because shale energy represents an opportunity as well as a challenge. Shale oil and natural gas present significant new energy resources, but their development also may pose risks to water quality and other water uses.[1] The intersection of evolving technology, environmental protection, hydrocarbon energy demand, and national and geopolitical energy and trade interests provide the context for this study.

Shale gas and shale oil[2] (collectively referred to as *shale energy*), which were long considered "unconventional" hydrocarbon resources, are now experiencing significant development in the United States. Shale oil and gas represent substantial fossil fuel resources for heating, electricity generation, transportation fuel, and industrial use. Economical extraction relies on directional drilling and hydraulic fracturing ("fracking"). This well-completion technique involves the injection of large volumes of water, along with water-conditioning chemicals and sand or other proppants, to pressurize and fracture shale formations to increase reservoir permeability.[3] The proppant holds the fracture open, allowing gas and oil to move to the well bore. A portion of the injected water, commonly referred to as "flowback," and naturally occurring water from the shale formation itself, referred to as "produced water," then return to the surface with the oil and/or gas.

For the purposes of this report, the combination of flowback and produced water, unless distinguished separately, will be referred to as "produced fluids." The term "wastewater" is also used, and includes produced fluids as described above, but may also contain other fluids produced during the drilling and development of shale energy wells.

The current level of freshwater used for fracturing and the management (reuse or disposal) of the produced fluids from the extraction are seen by some stakeholders as limiting factors in shale energy development. Shale energy development also poses the potential for contamination of surface water and groundwater resources through multiple pathways:

- accidental surface spills of chemicals used in hydraulic fracturing;
- accidental spill of wastewaters from well operations;
- improper disposal of wastewaters;

- well fluids leaking from valves and casings, including uncontrolled blowouts; and
- leakage and migration of gas and fluids at wells (e.g., improper well construction).[4]

While some of these concerns are specific to shale development, others are common to most energy development activities. However, the large volumes of fluids, chemicals, and injection pressures associated with high-volume hydraulic fracturing have posed new well development and wastewater management challenges for the industry and regulators.

This report discusses the water inputs to shale energy development, wastewater management related to shale energy development (including some related topics such as induced seismicity), and emerging water technologies for both the production of shale energy and the disposal of wastewaters. The report is intended to be a snapshot of current knowledge about water issues and technology development related to shale energy development. This report is limited to well development-related issues; it does not discuss water-related risks associated with transport of shale-derived energy resources.

PRIMER ON SHALE ENERGY RESOURCES DEVELOPMENT

The extraction of shale energy has the potential to affect U.S. energy security by reducing quantities of crude oil and refined petroleum products purchased in global markets. Additionally, shale gas contributes to the United States' effective independence regarding natural gas. When used as a fuel, natural gas, composed primarily of methane, is viewed as having lower emissions of many air pollutants relative to coal or oil as a fuel source, and is seen by some as a "bridge fuel" to a less greenhouse-gas-intensive energy future.[5] While these are some of the drivers behind interest in shale energy, how shale energy is developed bears directly upon its impact on water resources. This section provides a brief introduction to how shale energy is developed in the United States.

Location of Shale Resources

Shale energy deposits vary in size, depth, and quality across the United States. Some deposits occur primarily in one state—for example, the Barnett

Shale in Texas. Others underlie multiple states, such as the Marcellus Shale in New York, Pennsylvania, and West Virginia. The Marcellus Shale is currently considered the largest potential resource of shale gas, and is close to large energy-demand centers in the Northeast. The Utica Shale in Ohio and portions of the midwestern United States represents another sizable natural gas resource that is just starting to be developed. **Figure 1** illustrates key shale energy formations in the lower 48 states.[6]

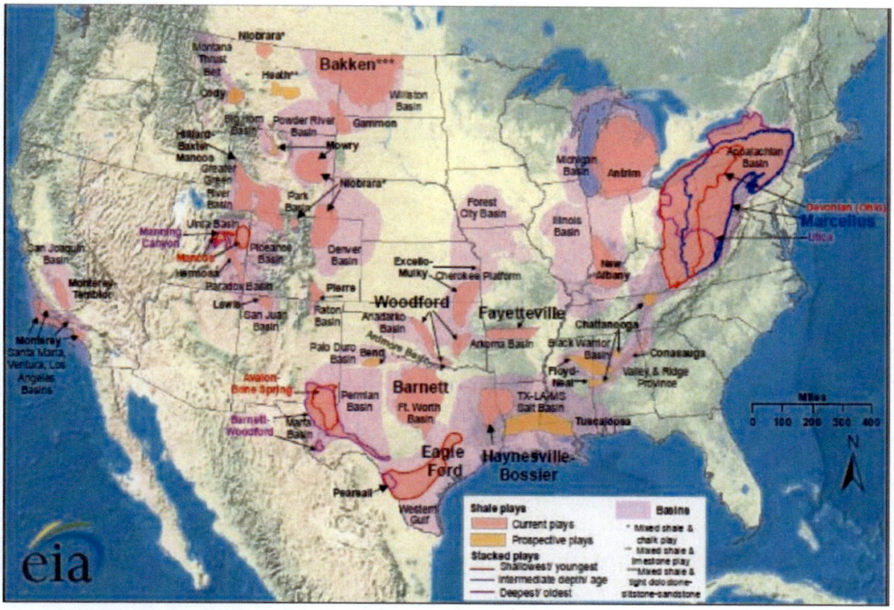

Source: U.S. Energy Information Administration, 2011.

Note: The term "play" does not have a specific definition, but generally refers to a set of known or postulated oil and/or gas accumulations sharing similar geologic and geographic properties, and containing a quantity of oil or gas that may be developed economically.

Figure 1. U.S. Shale Gas and Shale Oil Plays in the Contiguous United States.

Some deposits are primarily natural gas-bearing formations, while others contain significant oil resources. For example, well development in the Bakken Shale in North Dakota is often performed in pursuit of oil resources, while wells in the Eagle Ford Shale of Texas produce gas and oil in varying ratios depending largely on a well's location in the shale formation. Although interest in recovery of shale oil from the Monterey Shale of California continues, uncertainties remain about the near-term prospects for this development. Shale energy resources can typically be extracted economically only by using the hydraulic fracturing technique. This technique requires large volumes of water to pressurize the formations to increase reservoir permeability. Water is used in conjunction with a proppant for this purpose; proppant composition varies, but generally consists of sand or ceramic beads designed to be emplaced into fractures to maintain sufficient fracture width. The injected water opens up fractures and then delivers the proppant. Together, the injected mixture is known collectively as "fracking fluids."

To fracture a well, water is pumped from its storage site and mixed with the desired proportion of proppants and water-conditioning chemical additives.[7] Once blended, the mixture is injected into the well at pressures typically ranging from 8,000 pounds per square inch (PSI) to 10,000 PSI to achieve enhanced formation permeability. The volume of water, sand, and additives used for fracturing a horizontal well is typically about 4 million to 5 million gallons, but can vary from 2 million to 10 million or more gallons depending on the fracturing design and well type (e.g., fracturing of a vertical well often uses less water than fracturing a horizontal well).

In addition to commodity prices and market fundamentals of supply and demand, there are three key, interrelated sources of uncertainty affecting the pace of shale energy development: market structures, regulation, and public perception. Regarding market structure, shale gas and shale oil face some similar and some different uncertainties. Despite commonalities in current shale energy development technologies for shale oil and shale gas, the logistics of transporting oil versus natural gas to market are very different, as are pricing structures for the two commodities. Thus not all sources of uncertainty are likely to affect all segments of the shale energy industry uniformly.[8]

The second source of uncertainty is regulatory in nature. The policy attitude toward shale gas is evolving at multiple levels including local, state, and federal. Policy is evolving in the areas of air and water quality, water utilization, and land use (zoning). Substantial variations in regulatory approaches exist among states with active shale energy industries. Many such states, for example, permit some form of *forced pooling*,[9] which allows for horizontal drilling underneath a landowner's property (with compensation) even if the landowner has not explicitly signed a lease. Among active shale energy states, Pennsylvania and West Virginia do not have forced pooling in deep geologic formations (but do in shallower geologic formations from which oil and gas have been extracted for decades).[10] As of the end of 2013, Pennsylvania was in the process of developing policy that would permit forced pooling in the Marcellus Formation.

Regulatory uncertainty is challenging for shale energy production and transportation, particularly natural gas, due to the sunk nature of capital investments (power plants cannot quickly be repurposed, for example). Virtually all energy projects require large investments in capital that are sunk, but natural gas delivery is especially dependent on sunk capital, particularly pipelines and, where appropriate, liquefied natural gas (LNG) terminals. Since the mode of transportation for natural gas is not fungible (i.e., transportation cannot easily be shifted from one mode to another, as is the case with oil and coal), stable long-term supply contracts are generally required to encourage investment in gas transmission infrastructure in emerging gas shale plays (and in some shale oil plays where substantial flaring of natural gas occurs, such as the Bakken Formation in North Dakota), but such contracts are difficult to establish when future regulatory costs are unknown.

The third source of uncertainty is caused by substantial gaps between risks as understood and communicated by scientists; risks as communicated in media reports; and risks perceived by the general public. These gaps emphasize the importance of science-driven policymaking, and seem especially prominent in the case of risks to environmental resources, particularly drinking water quality. There is no systematic scientific consensus that hydraulic fracturing of deep shale formations, if done properly, poses threats to local drinking water supplies. Nevertheless, public perception in many areas is otherwise. Moreover, regulators have determined in various cases that shale energy well development and operations (separate from hydraulic fracturing) have impacted water quality.[11] Homes near drilling sites in southwestern Pennsylvania that rely on piped water systems have, on average, increased in value, while those that use on-site wells have, on average, declined in value. Similar evidence regarding public perceptions surrounding water quality issues has been gathered in the United Kingdom.[12] Several studies in 2011 and 2012 demonstrating some hydrologic connectivity between groundwater supplies and fracture zones in the Marcellus Formation[13] have been variously interpreted as suggesting an explicit link between drilling activities,[14] and suggesting exactly the opposite.[15] A 2013 study suggested a geospatial connection with drilling activities that may warrant further scientific and regulatory investigation.[16] The U.S. Environmental Protection Agency (EPA) is currently studying this issue, pursuant to a congressional request, but to date has not released any findings.[17] Gaps in scientific understanding on the potential impacts of shale energy development using high-volume hydraulic fracturing can heighten public concern and lead to increased regulatory scrutiny and uncertainty—note, for example, moratoria in Maryland, New York, and North Carolina.

WATER INPUTS INTO SHALE DEVELOPMENT

Hydraulic fracturing fluid must exhibit the proper viscosity and low friction pressure when pumped and used for well development. The fluid chemistry may be water-based, oil-based, or acid-based, depending on the properties of the formation. Water-based fluids, sometimes referred to as slickwater, are the most widely used, especially in shale formations because of their low cost, high performance, and ease of handling.

Water used in hydraulic fracturing may be piped or trucked from the source to the well-drilling area, depending on distance, rights-of-way, access, and topography. Water storage will typically occur on or near the well pad in either lined or earthen impoundments (typically built to codes determined at the state and local level), steel tanks, or temporary above-ground modular storage impoundments. The water is pumped from storage through a system of pipes, chemical blenders, pump trucks, valves, and pressure control devices (i.e., blowout preventers), and is then mixed with the desired proportion of proppants and chemical additives.

Fracturing initially requires significant water inputs, but while a well is producing there are few freshwater requirements unless refracturing is performed. Refracturing might be used to stimulate a well as production declines, possibly after a number of years. There are alternatives to water use for such procedures, but it is not known presently whether most shale energy wells will require refracturing or whether it will be economical to do so. Second, industry practices for water utilization, transportation, and treatment (or disposal) are evolving rapidly.

The following water sourcing topics for shale energy development are discussed below:

- water sources;
- costs associated with water inputs;
- water transport and storage; and
- access to water sources.

Water Sources: Availability of Groundwater and Surface Water

Regional differences in water availability may affect shale energy development over time. The most obvious constraints are likely to arise in arid to semiarid regions (but are not exclusive to arid and semiarid regions) that are already marginally to severely water-limited, and may become more so in the future.

The typical sources of water for well development have been surface or groundwater. Using surface water may require water transport to the well site via truck or pipeline, which may increase water-related costs and environmental and community impacts. Generally, surface water is a reliable water source in temperate regions, although it may become more difficult to access during drought periods or in isolated portions of a watershed. That is, although water may be regionally abundant in some regions, significant withdrawals can impact small streams in low flow periods. Groundwater can often be sourced at or near the drilling location, thereby reducing the costs and impacts associated with its transport for use in shale energy development. Water quality also must be considered for compatibility with hydraulic fracturing.

Shale energy projects in temperate regions, such as the Marcellus Shale and Utica Shale of the Appalachian Basin and the Haynesville Shale of East Texas and Western Louisiana, have used a combination of water purchased from municipal systems, industrial wastewater, and surface waters. In the Marcellus Shale in 2012, direct withdrawal from surface water represented 73% of shale energy water use; 27% of the water used came from municipal water systems.[18] In contrast, in the Eagle Ford Shale, aquifers have been the source for 90% of the water used in hydraulic fracturing; the other 10% is from surface supplies.[19] The following three figures illustrate where surface water and groundwater may be constrained given current levels of water use. **Figure 2** shows that the levels of use of existing surface water supplies (using precipitation as a simple measure of surface water availability) are already intensive in some locations. **Figure 3** shows that groundwater use in some locations exceeds aquifer recharge rates. For example, **Figure 3** shows that portions of Texas experiencing shale energy development like the Eagle Ford Shale (south Texas) and the Permian Shale (west Texas) had overdraft of aquifers at the onset of much of the shale energy development in 2005. **Figure 4** illustrates the variation in cumulative groundwater depletion over the course of more than a century for 40 U.S. aquifers.

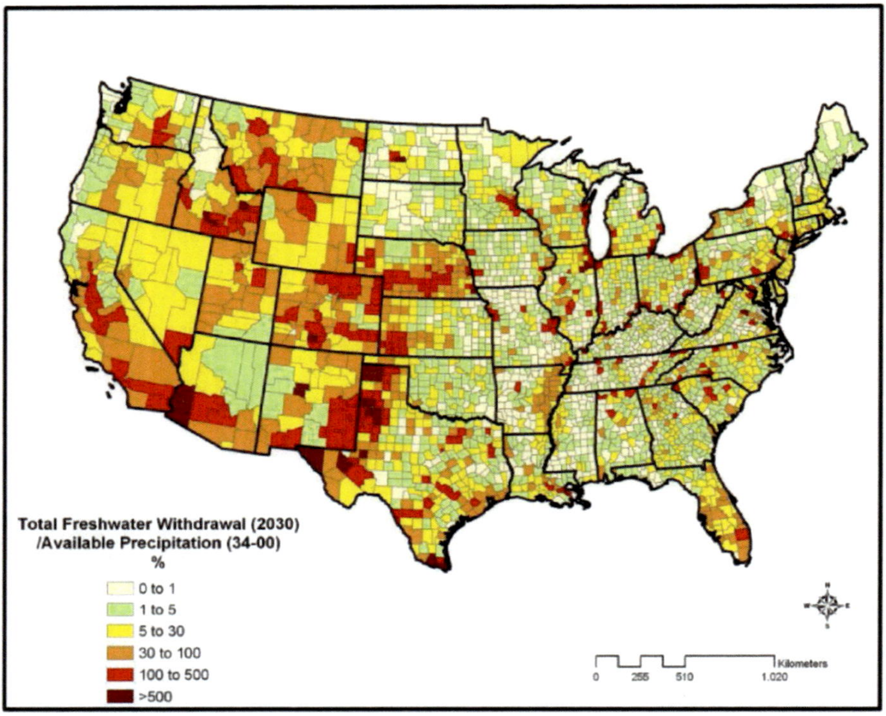

Source: Electric Power Research Institute, Water Use for Electricity Generation and Other Sectors: Recent Changes (1985-2005) and Future Projection (2005-2030), 2011 Technical Report, Palo Alto, CA, November 2011.

Notes: Higher values indicate the extent of water resource development in the area. Values greater than 100 indicate water imports from other counties and/or surface and groundwater storage. Data represent 2005.

Figure 2. Surface Water Use in the U.S. Contiguous United States; (withdrawal as percent of available precipitation).

Although groundwater and surface water supplies are the most common sources for shale energy development, there is interest in and some use of other supplies. The use of freshwater has raised concern that valuable water resources could be removed from the hydrologic cycle as a result of injection into shale, where the majority of injected water remains bound. Alternative sources of water such as treated industrial and municipal wastewaters or saline groundwater are often technically viable, and used to some degree. However, while the broader use of alternatives to surface water or groundwater is encouraged, economic, regulatory, legal, and technical conditions may limit their adoption.[20] The reuse of some wastewaters (e.g., abandoned mine drainage, produced fluids) as a substitute for freshwater may not only mitigate environmental damage or reduce disposal requirements for these wastewaters, but also reduce freshwater withdrawals. A concern, however, may be that those reusing these wastewaters may face liability risks under federal and state law. Several water source options and related issues are discussed below. Management practices from the American Petroleum Institute (API) stipulate that "whenever practicable operators should consider using non-potable water for drilling and hydraulic fracturing."[21]

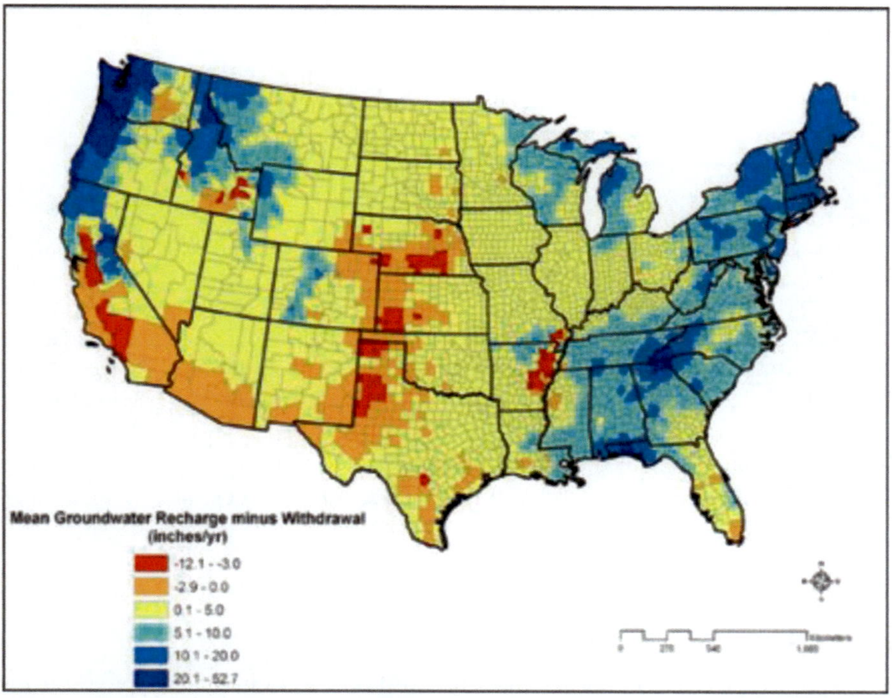

Source: Electric Power Research Institute, Water Use for Electricity Generation and Other Sectors: Recent Changes (1985-2005) and Future Projection (2005-2030), 2011 Technical Report, Palo Alto, CA, November 2011.

Notes: Negative values indicate that an aquifer is being mined at a rate that exceeds its recharge. Data represent 2005.

Figure 3. Groundwater Use in the Contiguous United States; (difference between recharge and withdrawal).

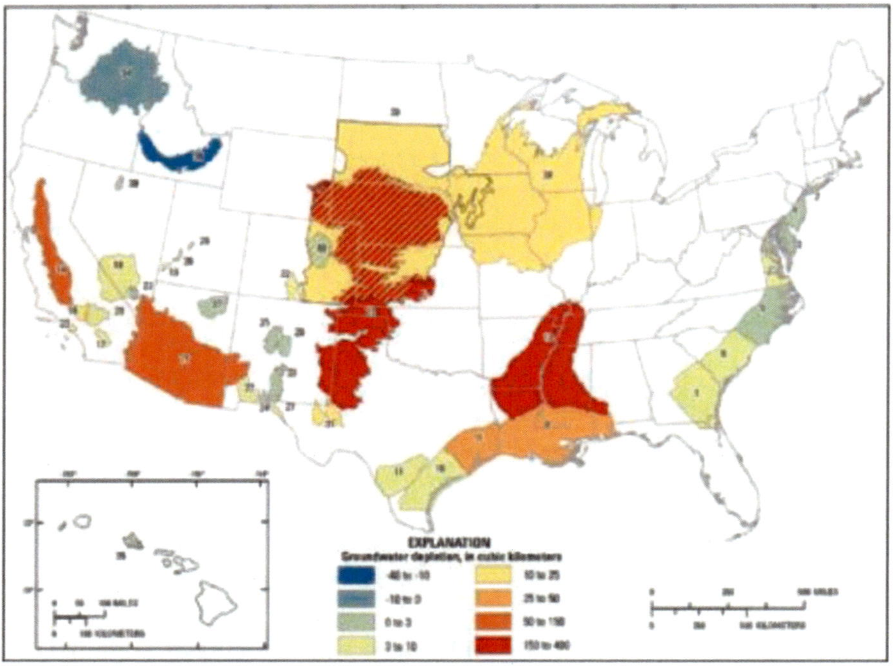

Source: Konikow, L., 2013, "Groundwater depletion in the United States (1900-2008)," U.S. Geological Survey Scientific Investigations Report 2013-5079, 63 p., http://pubs.usgs.gov/sir/2013/5079.

Note: Based on U.S. Geological Survey (USGS) studies of 40 selected aquifers. The USGS excluded Alaska from the map because no substantial groundwater depletion was evident in that state.

Figure 4. Cumulative Groundwater Depletion in the United States (1900 to 2008).

Recycled Produced Fluids

The recycling of produced fluids for well operation in hydraulic fracturing has been increasing over the last several years, especially in Pennsylvania. Based on a review of available Pennsylvania Department of Environmental Protection (PA DEP) records, produced fluid recycling increased from approximately 10% in 2009 to 90% by mid-2012. The advent of brine-tolerant friction reducers in slickwater fracking operations has allowed produced fluid reuse without compromising the effectiveness of well completions. Where produced fluids are being recycled in subsequent fracturing activities, the volume of recycled water may constitute between 10% and 30% of the total fluids composition. The Susquehanna River Basin Commission's (SRBC's) estimates indicate that approximately 70% of Marcellus wells recycled some produced fluids for hydraulic fracturing purposes. An increasing reuse trend is also occurring in other states such as Texas and Colorado, where water resources are scarcer. For example, in Colorado some operators indicate that all produced fluids were reused in hydraulic fracturing operations in the Piceance Basin.[22] In Texas, the percentage of produced water reuse varies by shale play, from 0% reuse in the Eagle Ford to 5% in the Barnett in 2011.[23]

Abandoned Mine Drainage

Use of discharge from abandoned mine drainage (AMD) in hydraulic fracturing applications has been limited. While AMD use has been employed by some operators where feasible, a number of technical, economic, and legal constraints have limited its use.[24] From a practical standpoint, the location of an AMD source must be sufficiently close to development activities to allow cost-effective transportation to well site(s), as shown in **Table 1**. Additionally, AMD chemistry must be carefully considered, as there is the potential for downhole precipitation of metals with sulfates that could cause fracture plugging and potentially impede gas flow and production. Therefore, the use of AMD, even from active discharge sites, may require treatment or at least significant dilution prior to use to minimize fracture-plugging potential.

Use of AMD for fracturing would face many challenges. Some abandoned mines have a clear line of ownership and liability for contamination of pristine waters with acidic mine drainage. In Pennsylvania, under the state's Clean Streams Law, waters from these mines must generally be treated to drinking-water quality before being released to streams. Other mines (referred to as "abandoned mines") do not have such a clear line of ownership and liability because the operators have ceased to exist. Waters from abandoned mines could potentially be captured for fracturing; however, potential liability under federal[25] and state laws likely would discourage use of AMD waters. Under Pennsylvania's Clean Streams Law, for example, any company engaged in the transportation and treatment of waters from abandoned mines would assume liability in the case of spillage or other infiltration into waterways. This is a potential disincentive for oil and gas operators to tap AMD from abandoned mines. One company (Seneca Resources), however, has begun a trial of limited AMD utilization in an area of northern Pennsylvania.[26]

Industrial and Municipal Wastewaters

Industrial wastewaters have the potential for use in fracturing operations where water is of compatible quality. Since each source of wastewater will have its own characteristics, these opportunities are evaluated on an individual basis. In addition, the location of the industrial wastewater with respect to drilling operations must be considered. Treated municipal wastewater may be used for well development (as well as having a number of other potential reuse applications outside of the oil and gas sector); this waste stream is typically treated to predictable levels that would be suitable for fracturing operations. As of 2012, there were three municipal treatment plants with the permitting approval to provide effluent to the shale gas industry in Pennsylvania. The Texas Commission on Environmental Quality reported approximately 30 industrial and municipal treatment facilities as of June 2012 that provide water to the industry.[27]

Costs Associated with Water Inputs

Water sourcing, transport, and storage practices utilized in shale energy development have been rapidly evolving to increase overall operational efficiency. Transportation from a source to a well site represents a substantial portion of water-related costs, as shown in **Table 1**; therefore, proximal source location with innovative water transfer methods increases cost-effectiveness.

Table 1. Per-Well Cost for Freshwater Sourcing and Transport (Marcellus Shale region)

Variable	Value
Volume required	11 million to 22 million liters
Per-unit water procurement costs	$1.25 to $5.00 per 1,000 liters
Truck transportation costs	$5.60/hour per 1,000 liters
Impoundment costs	$6.25 per 1,000 liters
Total cost for a single well	$13.80 to $17.75 per 1,000 liters

Source: Yoxtheimer et al., 2012, "The Decision to Utilize Acidic Coal-Mine Drainage for Hydraulic Fracturing of Unconventional Shale-Gas Wells," *Environmental Practice* 14:4, 7 p.

Water Transport and Storage

As shown above, transfer of water from source to site, as well as water storage, can be a significant operational cost. This section reviews common practices for the transportation of freshwater and produced water.

Water Intake Systems

A surface water withdrawal intake typically consists of a centrifugal or submersible pump installed at a stream, river, or lake withdrawal point that has been properly permitted. The intake structure itself must not be obstructive in order to avoid a water hazard. In addition, an intake screen must be utilized to prevent entrainment or impingement hazards for aquatic life. Oftentimes, the system is somewhat modular so that it can be moved for safety reasons such as during a severe flood. Some designs include an intake built into the streambed, which reduces sediment loading over time. Other designs are connected to a stream flow monitoring device to shut down the pump when flows go below permitted levels. Groundwater supply wells can also serve as intake systems, if properly constructed and designed to withdraw a volume of water needed to meet the demands of hydraulic fracturing operations. Challenges associated with the use of groundwater sources may include additional aquifer yield testing requirements for the purposes of permitting, and relatively low yields compared with surface water sources.

Water Trucking

Transportation of water and wastewater by truck represents a significant cost for shale energy water management. Total trucking costs, including fuel, are approximately $90 per hour.[28] Typical truck capacity is about 100 barrels (approximately 16,000 liters), so a well located one hour (round trip) from a freshwater source would require between 700 and 1,400 truck trips, representing $70,000 to $140,000 in transportation costs, or nearly $1 per barrel (roughly half a penny per liter). Thus, assuming all water is trucked in from a location that is a one-hour round trip from the well, water costs to develop a single unconventional shale well would be between $85,000 and $260,000, or between $1.21 to $1.84 per barrel of water ($13.80 to $17.75 per thousand liters). The transportation cost figure scales linearly with distance, while water cost is fixed; therefore, a well that is a two-hour round trip from a freshwater source would incur estimated costs of $2.21 to $2.84 per barrel of water ($27.60 to $35.50 per thousand liters), depending on the cost of water.

Water Pipelines

Direct piping of water from a source to a well-pad impoundment occurs in locations where operational costs are less than water transfer by truck and where pipelines can obviate the challenges and risks of transport by road. Use of pipelines for water transfer minimizes trucking, road damage, and diesel fuel use, and can be approximately 50% less expensive than trucking.[29]

Although initial capital costs are typically higher, the costs of installing permanent intakes, water pipelines, and impoundments may be recaptured if reused to serve multiple wells. A demonstration by Seneca Resources showed that an 11-kilometer pipeline and surface water withdrawal intake system would cost $7.2 million, but save about 50% ($9 million) in water transfer costs for fracturing operations at 70 wells.[30]

Centralized Water Storage Impoundments

Once delivered to the well site, water must be stored. Freshwater impoundment construction costs are approximately $1 per barrel, based on industry estimates,[31] which would equate to approximately $119,000 for a 19 million liter (5 million gallon) impoundment. This type of water storage method is fairly common, and can be cost-effective especially where long-term operations are anticipated; however, such impoundments cause significant earth disturbance.

Modular Water Storage

Use of temporary above-ground storage tanks and impoundments occurs more commonly because such structures reduce the surface footprint compared to centralized water storage impoundments. Further, the structures are reusable. An example is vertical steel tanks, which have the advantage of being capable of storing a large volume of fresh or produced water (up to 5 million gallons) in a relatively small area.

Access to Water Sources

The shale energy industry operates within a patchwork of local, state, and federal water management and regulatory regimes. In addition, regional organizations such as river basin commissions (RBCs), where present in areas with shale energy development, have emerged as active players in managing potential conflicts between watershed user groups including agriculture, energy, and public water interests.

State Approaches to Water Management

As shown in **Figure 5**, of 31 states surveyed, 30 regulate surface water and groundwater withdrawals through permits for water withdrawals or registration and reporting, and several states require both permits and registration and reporting.[32] Pennsylvania, for example, requires a water management plan (a full life cycle of the water used in shale gas production), although authority for most decisions is granted to the river basin commissions, except in the western part of the state, which lies outside the river basin commissions' boundaries. Louisiana, as another example, recommends that groundwater used for drilling or fracturing be taken from the Red River Alluvial aquifer. In Texas, surface water withdrawals for oil and gas rig operations require a permit; for groundwater, rig water supply does not generally require a state permit, but must comply with rules (e.g., registering wells, well spacing, well permit) of the respective groundwater conservation district. The rules are established by the districts, and vary widely. In North Dakota, the oil and gas industry accesses some of its water through water depots that are required to obtain relevant surface and groundwater permits. The state also can issue individual oil and gas operators' permits for access to surface and groundwater supplies. For some aquifers that are declining, North Dakota has limited access if other suitable sources are available.

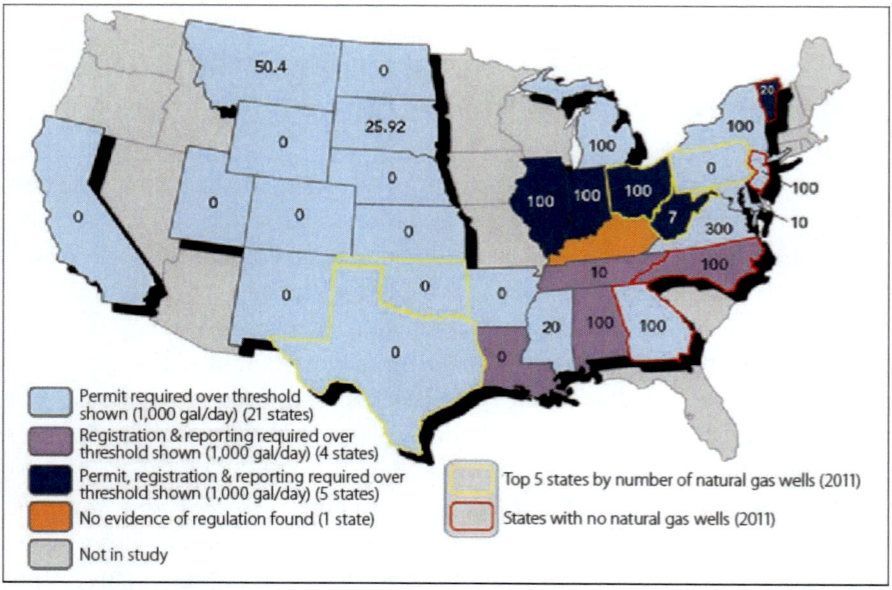

Source: Nathan Richardson et al., *The State of State Shale Gas Regulation*, Resources for the Future, May 2013, http://www.rff.org/centers/energy_economics_and_policy/Pages/Shale_Maps.aspx#map.

Note: Numbers shown on the states represent the withdrawal thresholds (in thousands of gallons per day) for reporting/permit requirements. Most apply to both groundwater and surface water withdrawals.

Figure 5. Lower 48 State Water Withdrawal Regulations.

River Basin Commissions

Relevant to shale energy development, a number of interstate commissions and compacts are important to the allocation of, and access to, freshwater—principally shared surface waters. Examples include the Susquehanna River Basin Commission (SRBC) and Delaware River Basin Commission (DRBC).[33] River basin commissions may have some authority to both ration water allocations among competing users and request member states to impose mandatory restrictions on "nonessential" water uses (e.g., golf course irrigation, lawn watering, service of water in restaurants, washing of most automobiles, etc.). They may regulate withdrawals (permit review), evaluate seasonal limitations, monitor water quality, tabulate water consumption and reuse, and establish moratoria on drilling to set limits and examine impacts, among other functions. River basin commissions in shale energy regions often have also expanded their research, monitoring, and staffing to meet the challenges of shale energy development.

WASTEWATER MANAGEMENT: FLOWBACK AND PRODUCED WATER (PRODUCED FLUIDS)

Once a well has been fractured and prior to coming on line, approximately 10% to 50% of the injected fluids may be returned to the surface over the course of several days to weeks, depending on the geology of the shale play.[34] These fluids are commonly known as flowback water, and consist primarily of the fluids used to fracture the shale formation. Flowback water is different from naturally occurring water in shale formations ("produced water") that typically is also brought to the surface following well completion. The produced formation water can be highly saline, and often is referred to as produced brines.

At some point, water recovered from a natural gas well will transition from mostly flowback water to mostly produced water.[35] In produced water, total dissolved solids (TDS) values range widely by shale play, from approximately 13,000 to more than 280,000 milligrams per liter (mg/L), with an average range of 13,000 to 120,000 mg/L, and can range as much as 120,000 to more than 280,000 mg/L within a play, as shown in **Table 2**. Produced waters may also contain constituents that are leached out from the shale formation, including barium, calcium, iron, and magnesium, as well as naturally occurring dissolved hydrocarbons and naturally occurring radioactive materials (NORM). Flowback water, shown in **Figure 6**, typically has elevated concentrations of TDS, which may include salts, metals, clays, and fracturing fluid chemical additives. The concentration of salts in flowback water increases rapidly during the first week or two after well completion. No clear demarcation exists between the two fluid flows. Proper storage and management of these fluids can prevent the potential contamination of groundwater and surface water that would occur if released into the environment. During the production phase of a well, some portion of the injected fluids in the shale formation may slowly flow out of the well as part of the produced water, along with natural gas or oil, typically at a rate of up to a few barrels per day, with the rate decreasing slowly over time (see **Figure 7**). The mixture of flowback and produced water is referred to in this report as "produced fluids."

Table 2. Salinity of Produced Water from Different U.S. Shale Formations

Shale Formation	Average TDS (PPM)	Maximum TDS (PPM)
Fayetteville	13,000	20,000
Woodford	30,000	40,000
Barnett	80,000	>150,000
Haynesville	110,000	>200,000
Marcellus	120,000	>280,000

Source: Acharya, H.A., Henderson, C., Matis, H., Kommepalli, H., Moore, B., Wang, H., 2011, *Cost Effective Recovery of Low-TDS Frac Flowback Water for Re-use*, U.S. Department of Energy: DE-FE0000784 Final Report.

Notes: TDS is total dissolved solids. PPM is parts per million (for reference, 10,000 ppm is equivalent to 1%). In this Department of Energy report, the authors refer to all returning water after hydraulic fracturing as "flowback," and do not differentiate between fracking fluid "flowback" and "produced water."

Practices for managing produced fluids vary widely by operator and by location. There is no identifiable set of best practices for water management for the shale gas or shale oil sectors as a whole. In Texas, for example, injection wells are widely utilized for wastewater disposal, whereas geologic disposal is utilized less frequently in the Appalachian region. Wastewaters from shale energy wells in the Appalachian region are more likely to be managed using a combination of underground injection, surface disposal (such as impoundments), onsite treatment and blending for reuse or transport to water treatment plants for reuse, surface discharge, or other disposal.

Source: Earth and Mineral Sciences Energy Institute, Pennsylvania State University.

Figure 6. Photograph of Flowback Water,Treated Flowback Water Ready for Reuse, and Produced Water.

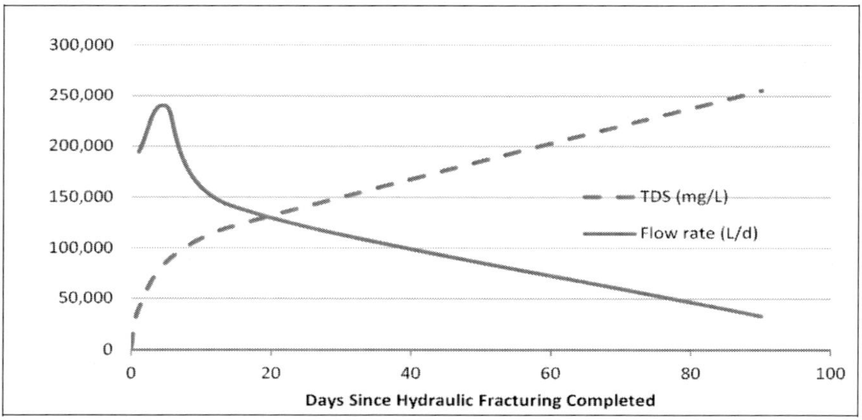

Source: Earth and Mineral Sciences Energy Institute, Pennsylvania State University.

Notes: Values on the Y axis reflect mg/L, which refers to milligrams per liter, for the TDS curve, and L/d, which refers to liters per day for the flow rate curve. For example, on day 20, where the curves intersect, the concentration was approximately 130,000 mg/L, and the flow rate was approximately 130,000 L/d.

Figure 7. Change in the Total Dissolved Solids Concentration of Produced Fluids over Time; (example of a Marcellus Shale well).

Surface disposal represents one of the lowest-cost ways to manage wastewaters from shale energy projects, but also introduces contamination pathways if impoundments are not properly constructed or managed. Hauling of water by truck to treatment facilities (or to geologic disposal wells if these wells are not located close to production areas) is among the highest-cost management strategies, and introduces potential contamination pathways if spills or other incidents occur during the transportation process.

Proper management of fluids derived from drilling and hydraulic fracturing operations remains a substantial environmental management challenge. Many operators have significantly improved their management of fluids by utilizing advances in technologies such as lining well pads to capture releases, using closed loop drilling systems, and recycling flowback and produced fluids. Continued improvement in fluids management practices is likely as companies further refine their operations to meet environmental challenges and regulatory requirements.

Table 3. Comparative Costs for Produced Fluid Management in Shale Energy Development

Treatment Method	$ per 1,000 gallons
Surface disposal	0.07
Deep injection well—existing	0.66
Evaporation/infiltration pond with spray	0.99
Spray irrigation	1.08
Microfiltration	1.36
Evaporative pond—lined-spray	1.97
Electrocoagulation	2.00
Shallow injection/aquifer renewal	2.85
Evaporative pond/infiltration	2.98
Water hauling	4.82
Deep injection well—new	5.64
Nano-filtration	6.15
Reverse osmosis	6.94
Evaporative pond—lined	27.56

Source: U.S. Department of Energy National Energy Technology Laboratory Project DE-FE0001466, 2012.

New treatment and reuse technologies are currently being deployed to further refine the treatment and recycling of flowback and produced fluid.

Deep underground injection wells (referred to as Class II wells under the federal Underground Injection Control (UIC) program)[36] are used to dispose of the portion of oil and gas wastewaters not recycled or sent to other locations for off-site treatment and disposal. Treatment and reuse technologies and practices for water sourcing, transport, and storage vary by operator and region. In addition, cost is always a consideration in fluid management practices. **Table 3** shows a range of costs associated with a variety of produced fluid treatment methods.

While disposal is a common management approach, others are seeking to identify ways to beneficially use these waste streams. The commercial or public-sector use of certain produced fluids, for example, is being permitted in Pennsylvania and West Virginia. That is, if the brines meet specified water quality requirements, they are being applied in winter to treat roadways in those states, rather than being disposed of geologically or treated in designated facilities.[37]

Identification of uses for waste materials that may be considered beneficial could be important to the process of designing regulatory frameworks that will allow drilling companies and potential users of materials that would otherwise be considered waste streams to make better decisions.

Underground Injection Disposal Wells

Deep well injection is regulated under the authority of the Safe Drinking Water Act Underground Injection Control (UIC) program, and is a common disposal method for a variety of waste fluids, including oil and gas wastes that are primarily produced waters (i.e., brines). Oil- and gas-related injection wells are classified as Class II injection wells. There are approximately 151,000 Class II injection wells in the United States, 80% of which are used for enhanced oil recovery and 20% of which are used for disposal of wastes.[38] Collectively, Class II wells accept an estimated 2 billion gallons of brine per day. The special class of oil and gas waste fluid disposal wells is collectively known as Class IId UIC wells. There are more than 30,000 such wells in the United States today, though the distribution of these wells among shale drilling areas is uneven. Texas hosts approximately 52,000 Class II injection wells, of which approximately 10,000 are disposal wells. Hence produced water recycling rates in Texas are generally less than 10%.[39] In contrast, the Appalachian Basin contains a limited number of Class IId injection wells, apparently due in part to the lack of suitable injection reservoirs with sufficient depth and permeability to accept significant volumes of waste, but also partly because of a lack of need for such disposal capacity before the emergence of shale gas. Through 2013, Ohio had approximately 180 active Class IId wells, while Pennsylvania had eight active disposal wells. Regulatory differences and policy issues also can play a role in well permitting.[40]

Example: The Marcellus Shale Play

Two questions important to the future development of shale energy resources in the Marcellus play, as well as other shale plays around the nation, are the following:

1) What is the volume of produced fluids projected to be generated over time?
2) What is the available long-term disposal capacity?

The available Pennsylvania Department of Environmental Protection (PA DEP) Marcellus gas and produced fluids records (from mid-2009 through mid-2012) were reviewed to evaluate this issue.[41] The volume of Marcellus Shale produced fluids and the associated portion disposed of via injection wells were compared to the available gas production records. Records from the first half of 2012 indicate that a total of eight barrels of produced fluids were generated for each million cubic feet of produced gas. Of the eight barrels of produced fluids, approximately 1.1 barrels were disposed of via injection wells.

Of the total produced fluids from all Pennsylvania oil and gas operations, 97% were disposed of in injection wells in Ohio.[42] **Figure 8** shows the change in volume between 1997 and 2012 of produced fluids injected into Ohio wells, indicating the rapid increase since about 2008. **Figure 9** shows the projected volume of fluids that may be generated from Marcellus Shale gas development in Pennsylvania based on trends from existing data, assuming a 5.2% annual increase in Marcellus gas production, as predicted by the U.S. Energy Information Administration (EIA).[43] Also shown in **Figure 9** is the current case of 1.1 barrels of produced fluids injected into UIC wells for each 1 million cubic feet of gas produced when 90% reuse is occurring. In addition, hypothetical scenarios are shown with an assumed 0.55 barrels and 2.2 barrels for each 1 million cubic feet of produced gas, and similar scenarios assuming only 2.6% year-over-year growth in Marcellus gas production.

Many factors can influence the volume and management of shale gas produced fluids. Lower-volume scenarios could result for a variety of reasons, such as low natural gas prices that might discourage the drilling of new wells and prevent previously drilled wells from being brought into production. New technologies may allow for more economical treatment and reuse of produced fluids, thus decreasing the percentage of the total amount produced sent for injection.

One significant unknown variable is the ultimate disposal capacity in Ohio, Pennsylvania, West Virginia, and New York. If it is assumed that 1.1 barrels of produced water are injected into disposal wells per million cubic feet of produced gas, then the volume of produced fluids requiring injection well disposal is projected to be 3.3 million barrels of Marcellus produced water (138 million gallons) from Pennsylvania alone by 2022. This would represent a 65% increase in injection well use as compared to the rates for the first half of 2012. The 179 injection wells used in Ohio (including 40 new wells brought on line over the last decade) were apparently able to handle an increase of approximately 6 million barrels (252 million gallons) annually over the course of a decade. In addition to possible limitations on capacity to inject all of the produced fluids if natural gas-related activities continue at the current pace or increase, the increased scrutiny on a possible link between injected fluids and earthquakes (discussed below) may also constrain the ability to install injection wells to handle all the disposal needs.

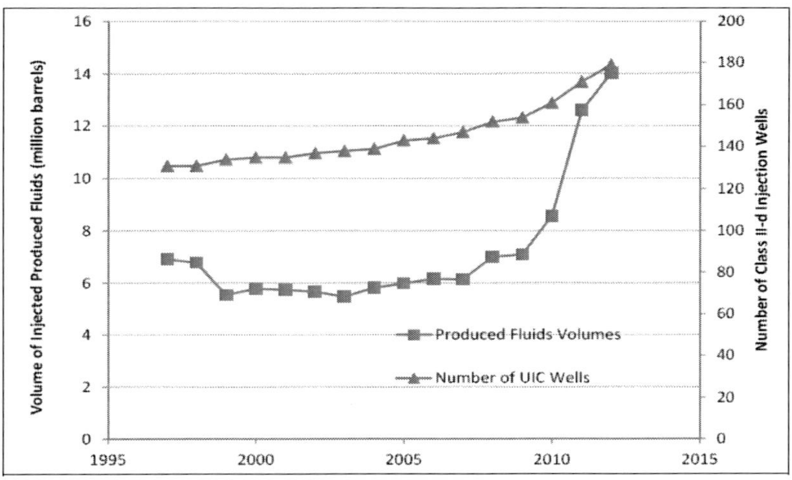

Source: Ohio Department of Natural Resources (OH DNR), 2012. Underground injection well data provided by Tom Tomastik, OH DNR Underground Injection Control Program Manager.

Figure 8. Ohio's UIC Disposal Well Activity, 1997-2012; (volume of produced fluids injected and number of UIC wells).

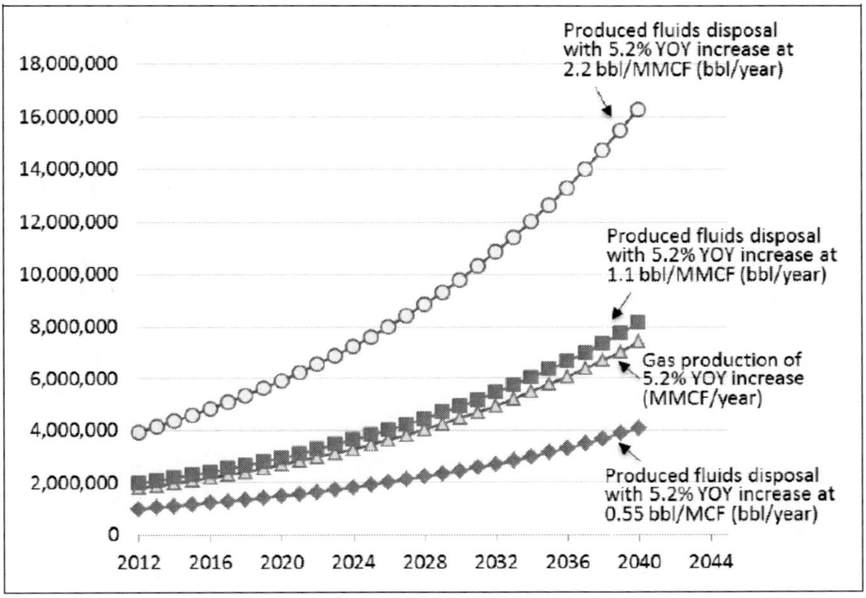

Source: Earth and Mineral Sciences Energy Institute, Pennsylvania State University.

Notes: Based on trends from existing data, as well as projected increases of 2.6% and 5.2% annually in Marcellus gas production. YOY means year-over-year.

Figure 9. Projected Fluids from Marcellus Shale Gas Development in Pennsylvania.

RELATED ISSUES: INDUCED SEISMICITY AND ABANDONED WELLS

Potential for Induced Seismicity

Induced seismicity[44] is not a concern related to surface or groundwater resources, per se, but has been raised as a potential issue. While fracturing itself involves induced seismicity, such events are localized and of very low amplitude (10^{-2} on Richter scale); they generally cannot be felt at the surface. Reports of minor earthquakes possibly induced by fracturing occurred in Garvin County, OK, in 2011,[45] but no definitive connection to fracturing per se has been made. There is, however, a potential for induced seismicity anywhere that wastewater is pumped into deep rock units at high rates,[46] regardless of regional geologic contrasts.[47] One theory suggests that "fluid injection may trigger earthquakes if pressures, rates, and permeability are sufficient to allow fluid to reach a favorably oriented fault and reduce the normal stress, decreasing fault strength."[48] The potential depends on a number of factors, including (1) the state of subsurface stresses (i.e., whether stress buildup has been relieved by previous earthquakes); (2) the presence or absence of through-going faults; (3) porosity and permeability (transmissivity of fluids) of the unit into which fluids are being pumped; and (4) the rate at which fluids are being pumped and the relative pressure differential developed. It is likely that induced seismicity has occurred in what are generally considered "stable" tectonic regions (compared to, for example, portions of California), including eastern Ohio, Oklahoma, and Arkansas.[49]

Earthquakes with magnitudes as high as 4.8 have been measured in some regions where the injection of wastewater from drilling/completion activities occurs. One example is a series of earthquakes in the Dallas-Ft. Worth area (Barnett Shale) that has been linked to underground injection wells.[50] In 2011, a series of low-magnitude earthquakes (magnitude 2.1 to 4.0) were recorded in the Youngstown, OH, area. This seismic episode was not itself caused by fracturing, but was linked to the operation of a UIC Class II disposal well that was used to dispose of wastewater from Marcellus Shale drilling in Pennsylvania. Following the incident, the disposal well was shut down.[51]

Abandoned/Orphaned Wells

Abandoned oil and gas wells are a concern in some areas of active shale energy extraction, because intersection with previously unmapped wells provides a potential pathway for migration of methane or fluids into groundwater. In particular, "orphaned" wells, those that are defined as inactive/abandoned oil and gas wells with no responsible party to properly plug the well and restore the location, are of concern because their location and status are often unknown. Abandoned wells must be plugged to permanently seal the inside of the well and wellbore (typically above and within producing zones and across freshwater aquifers) so that fluid cannot migrate from deeper to shallower zones or create reservoir problems through downward drainage. The plugging process involves the placement of cement and other materials, such as gels or bentonitic mud, within the wellbore and production casing in a manner that prevents the upward or downward migration of formation fluids. All oil and gas producing states now regulate well plugging; most have standards for cement quality, and most require advance notice so that regulatory personnel can witness operations to assure proper plugging.[52]

Regional Concern in the Marcellus Shale and Utica Shale Plays

In addition to the sheer number of abandoned wells, many such wells were drilled prior to requirements for regulation, permitting, and record-keeping, or when those requirements were less stringent than current requirements. For example, the first well in Pennsylvania was drilled in 1859, and the first requirement to plug was issued in the 1890s. Pennsylvania first imposed regulation of oil and gas wells in 1955. The state issued permits to drill through coal seams in 1956, and for the drilling of all wells in 1963. The Pennsylvania Oil and Gas Act of 1984, which took effect in 1985, required the registration of all wells that were not previously permitted. Nonproducing wells were required to obtain an inactive regulatory status or be plugged. Pennsylvania has documented nearly 34,000 preregulatory wells, but estimates suggest that the total number may approach 200,000 wells (see **Figure 10**).

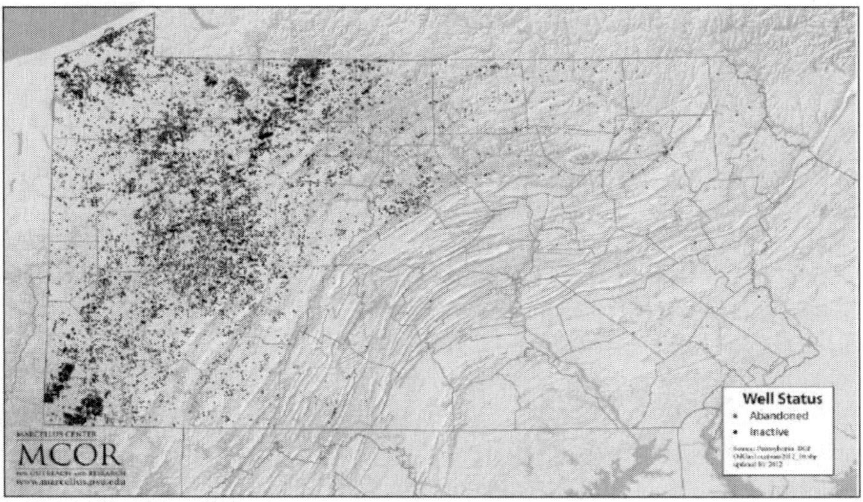

Source: Marcellus Center for Outreach and Research, based on Pennsylvania Department of Environmental Protection data as of October 12, 2012.

Note: Approximately 9,000 wells shown; as many as 200,000 additional wells may exist.

Figure 10. Known Abandoned and Inactive Wells in Central and Western Pennsylvania.

Wells that may not have been properly plugged and cased can be a source of methane migration from gas-bearing strata at somewhat greater depth to the surface and/or into freshwater aquifers. A number of abandoned wells in Pennsylvania penetrate rock strata to the Oriskany Sandstone below the Marcellus Shale. Numerous instances of methane migration to private water wells have been linked to nearby abandoned wells.[53] Such wells, when unknown, are also a potential danger when further drilling occurs nearby. A previously unmapped well blew a geyser of methane gas and water up to 30 feet in the air in June 2012 in Tioga County, PA (in the northeastern part of the state), during drilling of a Marcellus Shale well.[54] Orphan wells likely contribute significantly to the flux of methane to the atmosphere, providing an additional, untallied source of greenhouse gases. A more detailed discussion of the relative contribution of greenhouse gases to the atmosphere from orphaned wells is beyond the scope of this report.

Responsibility for the management of abandoned and orphaned wells typically falls to state authorities. In Pennsylvania, for example, wells fall under the jurisdiction of the Pennsylvania Department of Environmental Protection (PA DEP). Under that state's well plugging program, 2,948 wells had been plugged through 2013.

EMERGING WATER TECHNOLOGIES FOR SHALE ENERGY DEVELOPMENT

The pace of technological change in water sourcing and water management for shale energy development is rapid, but uneven. Trends in water management have generally been influenced by local disposal costs, regulations, and geologic conditions, rather than by water scarcity alone. Some regions, particularly those where regulations restrict the discharge of wastewater to surface waters, and which have relatively few options for wastewater disposal (due to a combination of geologic and regulatory factors), have seen shifts toward closed-loop water management systems that utilize recycled flowback water extensively and minimize the use of disposal wells. These systems have also been used more extensively, and by necessity (because of a lack of wastewater injection wells), in emerging unconventional production areas such as the Marcellus Shale play than in regions with recent growth in shale development, but that have a long history of active oil and gas production, such as Texas.

This section discusses the status of emerging technology options for reducing the potential impacts of shale energy activities on groundwater and surface water resources. Much of the research is being conducted by private industry, often in close partnership with government agencies and university scientists.

Technology Options for Drilling and Completing Wells

At the drilling, completion, and production phases of the shale energy well life cycle, a number of alternatives to conventional water-utilization systems are being implemented. These include

- *nontoxic or "green" fracturing fluid additives*, driven in part by concerns over the composition of fracturing fluids and increasing requirements of disclosure of fracturing fluid composition;
- *alternatives to freshwater in the fracturing process*, including recycled flowback fluids (mixed at various proportions with freshwater), carbon dioxide, nitrogen, hydrocarbon gases (such as ethane and propane), industrial waters, and (in the Appalachian region) potentially acid mine discharge waters;
- *innovative well and well-pad configurations* such as multilateral wells, which, in some cases, reduce the total volume of fluids required, but are more likely to have economic advantages in reducing labor, trucking, and other water handling costs; and
- *closed-loop or reduced emission ("green") well completions* for handling flowback fluids and minimizing the venting of methane to the atmosphere.

These innovations are at various stages of maturity. Continued deployment of these innovations may be driven by a mix of project economics and regulatory influences (such as regulations regarding closed-loop completion systems that will be required for many shale energy projects beginning in 2015).[55] Some of these innovations, such as closed-loop completions, are becoming commonplace in many producing regions, while others, such as nontoxic additives and alternative fracking fluids, need additional demonstration and validation before being accepted more broadly by industry.[56]

Nontoxic Hydraulic Fracturing Fluid Additives

New additives and transparent reporting of chemical additives to hydraulic fracturing fluids,[57] regardless of their toxicity, are already being applied in some cases. Companies such as Halliburton, Schlumberger, Baker-Hughes, and others already have them available and are continuing to develop new additives, according to their individual reports on websites and investor circulars.[58] In Pennsylvania, industry is now required to report types and volumes of additives on the FracFocus website.[59] Pennsylvania joins Texas, Colorado, Arkansas, Montana, Michigan, and other states requiring some level of disclosure of volume or composition of fracking fluids, or both. This trend toward nontoxic additives (referred to as "green" fracking fluids) has the potential to provide greater protection for workers and lowered impact of spills on surface waters, soils, and shallow groundwater.[60]

Federal law does not require disclosure of the chemical composition of hydraulic fracturing fluids. The U.S. Department of the Interior has proposed rules requiring disclosure for wells drilled on public lands.[61] All states with chemical disclosure requirements provide various exemptions for proprietary chemicals that are considered "trade secrets" specific to a particular company.[62]

A few examples of additives, which have specific purposes, are

- biocides to prevent bacterial growth that could inhibit well performance and possibly create potentially toxic gases such as hydrogen sulfide;
- friction reducers to minimize the power needed to pump hydraulic fracturing fluids downhole to create the level of pressures required for effective fracturing of reservoir rock; and
- scale inhibitors to prevent minerals from precipitating at critical places in a well that might significantly reduce production efficiency.

Some traditional additives are toxic, and can reappear in flowback water. A number of large industry players have committed to eliminating some additives by conducting tests of their effectiveness in different formations and/or to providing suitable nontoxic substitutes that are effective during reservoir stimulation.[63] Some of these alternative additives were originally developed for use in the food industry.[64] To be successful in the marketplace, the performance of such "green" additives must equal or exceed the performance of traditional fracture stimulation fluids.

Alternative Hydraulic Fracturing Fluids and Methods

A potential water-saving process with other potential advantages is the use of carbon dioxide (CO_2) or other water-free agents such as nitrogen (N_2), methane (CH_4), ethane (C_2H_6), propane (C_3H_8), and butane (C_4H_{10}), as fracturing fluids. Nitrogen is a common inert, nonsorbing, and compressible fracturing fluid (usually used as a foam); carbon dioxide is a corrosive, highly sorbing,[65] compressible fluid; and methane, ethane, and propane are noncorrosive, highly sorbing, compressible fluids. This category or grouping of gases is often referred to as liquefied petroleum gas (LPG). An additional advantage to their use is that these alternative fluids may limit formation damage that characterizes the application of water to certain shale mineralogies, particularly those rich in certain clay minerals that swell upon contact with water.

A fundamental motivation for the use of carbon dioxide is the possibility for superior performance in generating the connected pores that allow a more efficient extraction of natural gas, essentially increasing the permeability so that natural gas can flow more easily from the pores in the rock to the production well. In addition, as illustrated in **Figure 11**, the use of carbon dioxide may enhance production of methane because carbon dioxide can replace or "kick off" methane sorbed to the solid organic material in the shale. This property could also allow for a modest sequestration of carbon dioxide in a shale reservoir.

The LPG combination is sometimes referred to as "gas frac" methodology. It has been applied successfully in tight-gas sand reservoir stimulations, primarily in Canada.[66] In addition to reducing water use, flowback, and formation damage, there are additional benefits to each of the alternative fluids, and some disadvantages. One advantage is that all of the gas flowback after stimulation can be recaptured at the wellhead and reused. Also, for some shale formations, the use of hydrocarbon gases prevents "water blocks" (in which water clogs pores in low-permeability shale formations) that might occur with slickwater fracking. Carbon dioxide provides the same benefit.

One disadvantage of using carbon dioxide and other gases instead of water is the relatively high commodity costs, as well as transportation costs for linking sources of carbon dioxide and other alternatives to a well site. Their use also may raise a number of other issues related to safety and possible environmental impacts. If LPGs are used instead of water, first responders or emergency personnel may be exposed to additional risks in the case of well fires, blowouts, or other incidents. In addition, using LPG fluids introduces the possibility of fugitive hydrocarbon emissions during or after the completion of a gas frac, which could pose health and environmental concerns for groundwater, surface water, and air quality.

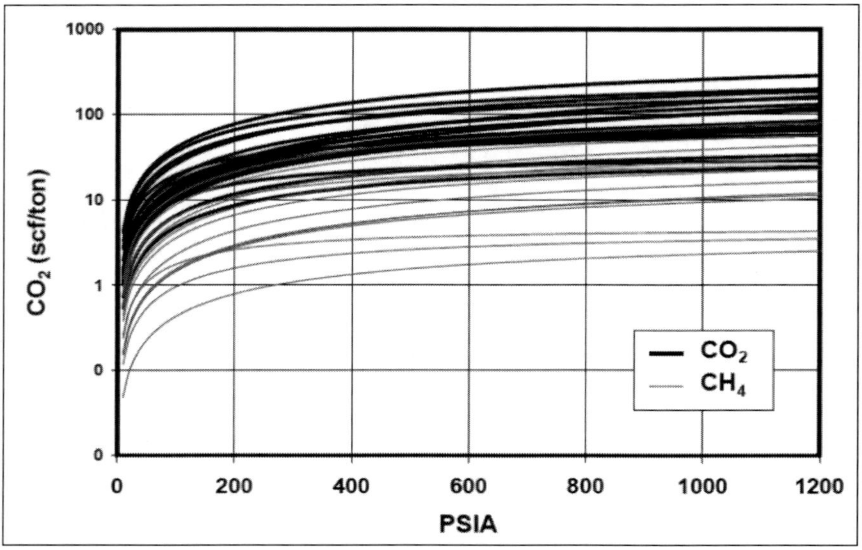

Source: Nuttall, B.C., Eble, C.F., Drahovzal, J.A., and Bustin, R.M., 2005, *Analysis of black shales in Kentucky for potential carbon dioxide sequestration and enhanced natural gas production*, Final Report to U.S. Department of Energy, DE-FC26-02NT41442.

Note: The figure provides a summary of adsorption isotherms (where psia measures gas pressure) and indicates a higher sorption capacity (Y axis) for carbon dioxide compared to methane, which means that carbon dioxide would be preferentially adsorbed to black shale and methane would be released.

Figure 11. Carbon Dioxide (CO_2) and Methane (CH_4) Adsorbtion onto Organic-Rich Devonian Black Shale From Kentucky and Ohio; (as a function of gas pressure).

Multilateral Wells

This technique, using lateral horizontal wells that branch off the main vertical wellbore, so that multiple shale horizons can be tapped from a single surface well pad, often leads to a reduced surface footprint and improved economics. It does not necessarily lead to savings in water volume used during fracturing. Other advantages, however, include less rig time, truck traffic, and fewer fluid lines.[67]

Closed System Completions

Closed-loop systems (referred to as reduced emission or "green" completions) for handling flowback and reducing gas leakage and flaring have been used in some U.S. gas shale plays. Closed-loop systems help to minimize the exposure of produced fluids to the environment (air or water), with the intent of reducing the risk of water contamination and air pollution.

Typically, in regions of rapid hydrocarbon exploration, the rate of well drilling exceeds the ability of industry to bring gathering lines (small-diameter pipelines to provide takeaway capacity for natural gas) to individual well pads. When completing wells that are fractured without gathering lines in place, there is a period of three to 10 days (up to 60 days) during which produced fluids from the well must be captured, stored, and ultimately disposed of, or treated or reused. During this period, natural gas also flows from the well, but cannot be effectively captured without storage and transport facilities in place. It is a common practice to "flare" the gas—burning the produced gas and converting it to carbon dioxide—rather than venting natural gas directly (carbon dioxide is a less powerful, albeit more persistent, greenhouse gas than methane). One company, Devon Energy, has dedicated itself to using such closed-loop completions in the Texas Barnett Shale. The Barnett Shale play, however, has the advantage of an existing oil and gas production infrastructure in a well-established producing area.[68] In newer shale plays such as the Marcellus Shale and the Bakken Shale, wells may be drilled prior to the development of the infrastructure needed to transport gas to market (see **Figure 12**). In such instances the natural gas generated during well completion is typically flared. For example, from 2008-2012 gas production in North Dakota from the Bakken Shale oil play accounted for 0.5% of total natural gas extracted in the United States; however, the amount flared in North Dakota was approximately 22% of all natural gas that was either flared or vented in the United States.[69]

The EPA has mandated that, with some exceptions, onshore natural gas wells must adhere to "green completion" guidelines by 2015.[70] This means that completions must be made within a closed system that allows separation of the water and gas phases, thereby significantly reducing greenhouse gas emissions, as well as those of volatile organic compounds (VOCs).[71] (The EPA rule does not apply to wells drilled primarily for production of crude oil, such as wells in the Bakken Formation.)

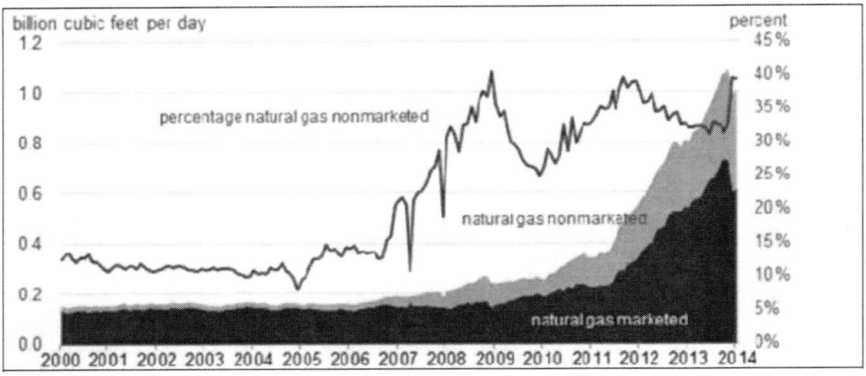

Source: U.S. Energy Information Agency, *Nonmarketed Natural Gas in North Dakota Still Rising Due to Higher Total Production*, 2013, http://www.eia.gov/todayinenergy/detail.cfm?id=4030, and http://www.eia.gov/todayinenergy/ detail.cfm?id=15511&src.

Notes: Natural gas production in North Dakota's portion of the Bakken Formation has grown with increased oil production. In 2013, natural gas production continued to outpace pipeline capacity: nonmarketed natural gas increased to an average of 0.13 billion cubic feet per day (Bcf/d) through the end of 2013, compared to 0.16 Bcf/d levels in 2011. However, nonmarketed gas, as a percentage of total production, decreased from 37% in 2011 to 33% in 2013, as several infrastructure projects came online. Most nonmarketed gas is flared.

Figure 12. North Dakota Natural Gas Production; (marketed and nonmarketed gas 2000-2013).

Produced Fluids Management and Treatment Technologies

A variety of produced fluids water management strategies and treatment technologies are being used in shale energy development to reduce the need for use of freshwater and disposal of produced fluids. Treatment costs can vary widely by method, as outlined in **Table 3**, from a few cents to tens of dollars per thousand gallons treated. Advances in new water treatment technologies are being developed domestically in response to evolving demands of the shale energy industry, and also are being imported from an array of international and foreign companies with specialized expertise.

Produced Fluid Treatment and Recycling Technologies

The recycling of produced fluids is increasing in shale plays across the United States, most prominently in the Marcellus Shale play. The primary driver for water treatment prior to reuse of produced fluids for hydraulic fracturing operations is to minimize the possibility of shale gas reservoir damage, such as chemical or physical plugging, that might be induced by constituents present in produced fluids. A damaged reservoir could reduce oil or gas production. In particular, high chloride levels can interfere with friction reducers and reduce fracturing efficiency, while divalent cations such as barium, strontium, calcium, and iron can precipitate with sulfates or carbonates, thus forming scale within fractures and contributing to fracture plugging.

As shale energy development and produced fluids reuse for fracturing operations have increased, operators have increased their use of a suite of treatment technologies to minimize the potential for shale reservoir damage. The increased reuse is due in part to improved fracturing mixtures that are brine-tolerant, thus allowing the use of produced fluids for hydraulic fracturing. Based on review and analysis of Pennsylvania Department of Environmental Protection (PA DEP) records for unconventional well development for 2012, 23.2 million barrels of produced fluids were reused out of a total of 26.8 million barrels generated, a reuse rate of approximately 87%. By comparison, the reuse rate in Pennsylvania in 2011 was 72% (12.1 million barrels reused versus a total of 16.9 million barrels of produced fluids). The percentage of reuse varies in other states. In Colorado, some reports indicate that most produced water is reused, and some operators claim that all produced fluids were reused in hydraulic fracturing operations in the Piceance Basin.[72] The percentage of reuse varies by shale play in Texas, but appears to be much lower than in Colorado, from 0% reuse in the Eagle Ford Shale play to 5% in the Barnett Shale play, based on information from 2011.[73]

There are two major recycling approaches: use of field management technologies deployed at or near drilling sites, and use of centralized treatment facilities, as described below.

Field Treatment and Recycling

A variety of approaches have been developed to reuse produced fluids in the field, with the primary advantages of minimizing the transport of wastewater, which reduces trucking costs, fuel use, carbon emissions, the potential for trucking accidents, and road damage. The major requirements by operators for the use of these technologies are that they effectively remove contaminants, have high recovery rates, are low maintenance, have a small footprint, and are operationally robust enough to handle a range of fluid qualities. A review of the options for field treatment and reuse, along with advantages and disadvantages, is summarized below.

Direct Reuse with Blending

Recovery of produced fluids and direct reuse of them for subsequent hydraulic fracturing typically involves blending of the return fluids with fresh makeup water in order to have the necessary volume of water for hydraulic fracturing. This approach may involve allowing coarser sediments to settle out in tanks; however, suspended particles may remain.

The primary advantage of this technique is the relatively low costs involved with storage of fluids in approved containment (e.g., double-lined centralized impoundments or steel tanks) and the operational costs associated with blending in freshwater. However, the quality of such blended water may be suboptimal. A disadvantage with direct reuse is increased risk of reservoir damage associated with either suspended sediments or multivalent scaling agents such as calcium, barium, strontium, iron, sulfate, or carbonate.

Filtration

Filtration technologies range from the use of bag filters designed to reduce suspended sediment concentrations to more sophisticated micro- or nanofiltration technologies with the ability to also reduce multivalent ion concentrations (scalants).

Based on a survey of Marcellus Shale play operators, the industry criteria for produced fluids reuse are shown in **Table 4**, including suspended particle size of <20 micron,[74] which can be achieved by all advanced filtration technologies. Filter socks would not reduce scaling agent concentrations, but micro- or nano-filtration would be effective in ion removal (although this would require power and additional operational oversight, thus increasing the cost).

The advantage of filtration technologies is that they require low to moderate maintenance while achieving moderate to high scalant removal efficiency. These technologies also achieve high recovery (>90%); therefore, they have high reuse potential, thus minimizing the need to dispose of residual wastes. Waste consisting of either spent bag filters or reject waters requires appropriate disposal, and adds to waste management costs.

Table 4. Suggested Maximum Concentration of Chemical Constituents in Produced Fluids for Reuse

Chemical Parameter	Maximum Value (mg/L)
TDS	50,000
Hardness	26,000
HCO3	300
SO4	50
Cl	45,000
Ca	36,000
Na	8,000
Mg	1,200
K	1,000
Fe	10
Ba	10
Sr	10
Mn	10

Source: U.S. Department of Energy National Energy Technology Laboratory, RPSEA Project 07122-12, 2009, *An Integrated Framework for Treatment and Management of Produced Water.*

Note: mg/L is milligrams per liter.

Chemical Precipitation

This class of treatment technologies uses a relatively conventional chemical addition process to remove scalants from the wastewater stream by increasing the pH and adding a coagulant that causes positively charged ions (cations) to precipitate out as sludge. The water is then run through a clarifier, and the sludge is separated, collected, dewatered, and ultimately disposed of in a permitted landfill. This process is highly effective at removing scaling agents at a moderate cost, though it does require greater maintenance to adjust chemistry with varying influent water quality. **Figure 13** shows the typical treatment scheme for use of this technology.

Electrocoagulation

Electrocoagulation is the process of destabilizing suspended, emulsified, or dissolved contaminants in an aqueous medium by introducing an electric current into the medium through an electrolytic cell with one anode and one cathode. Once charged, the particles coagulate to form a mass, and can be combined with electroflotation to effectively remove contaminants from water with the advantages of reduced sludge production, no requirement for chemical use, and ease of operation, with recovery rates of approximately 95%.[75]

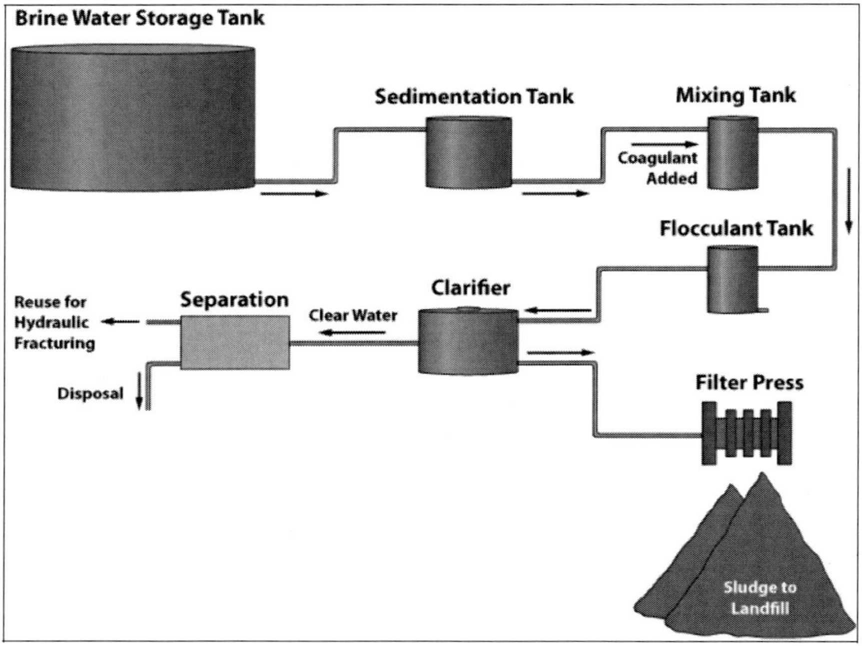

Source: Earth and Mineral Sciences Energy Institute, Pennsylvania State University.

Figure 13. Typical Chemical Precipitation Treatment Scheme for Produced Waters Reuse.

Desalination

Mobile desalination technologies have been developed to remove a high percentage of total dissolved solids, including both scaling agents and salts. The most widely used technologies include pressure-driven (i.e., reverse osmosis) and thermal-driven (direct heat), or a combination of pressure- and thermal-driven (mechanical vapor recompression, or MVR) technologies. The advantage of desalination is that a very clean effluent is produced and can easily be recycled, or, with proper permitting, even potentially discharged to a stream or river. The primary disadvantage is that the electricity required results in high associated energy costs.

Reverse osmosis can be used to treat only fluids having a total dissolved solids (TDS) of approximately less than 45,000 milligrams per liter (mg/L), and therefore can be used only in shale plays with lower-TDS produced fluids. It may be most effective in the Fayetteville Shale or Woodford Shale, but not the Marcellus Shale.[76] Pretreatment is generally required (typically chemical precipitation) to avoid membrane fouling. In addition to producing treated water, reverse osmosis also produces an even more saline waste concentrate which requires handling and disposal.

In contrast, thermal technologies can handle TDS loads of 100,000 mg/L (or higher), and therefore can be broadly applied in most shale plays. Thermal processes require pretreatment to soften water prior to either application of direct heat to boil the water, or use of MVR, where the water is both heated and compressed to add the energy required to boil water. In MVR the heated water is fed through preheat exchangers to absorb heat from the distillate and concentrate products, and passes into a recirculation loop where concentrate circulates through an evaporator exchanger and a vapor/ liquid separator.[77] Fluid recovery using direct-heat thermal technology results in fluid recovery efficiency of upwards of 56%,[78] whereas use of MVR is more energy-efficient and can achieve fluid recovery efficiency of upwards of 90% efficiency for reuse.[79] The higher efficiency means less concentrate to dispose of or treat further.

Centralized Treatment Facilities

The use of centralized treatment facilities for produced fluids management involves the use of a similar suite of technologies as summarized above. Whether centralized or on-site treatment is a preferred option depends on the trade-off between the cost of transporting produced fluids to and from the treatment site and the economies of scale possible with larger treatment facilities.

Depending on the location of the facility in relation to drilling locations, transport distances can be great; therefore, trucking costs can be significant.

As indicated by PA DEP records, during 2012 there were 17 treatment facilities in Pennsylvania that actively treated Marcellus Shale wastewater for reuse for hydraulic fracturing. All of the facilities relied on chemical precipitation as the primary treatment, with two other facilities having advanced desalination capacity using thermal technologies. The total capacity of the facilities was approximately 4 million gallons per day. In contrast, based on analysis of Marcellus Shale wastewater during 2012, on average only about 15% of the waste (462,000 gallons per day) went to centralized treatment facilities for recycling purposes. This suggests that approximately 11% of the existing recycling treatment plant capacity was utilized. The remaining 85%

(2.6 million gallons per day) of the recycled produced fluids was managed in the field.

While the most recent PA DEP waste production data suggest an estimated 87% of produced fluids were being recycled for hydraulic fracturing operations, approximately 13% of the fluids needed to be disposed of according to applicable regulations. The primary means of disposal of the remainder of the produced fluids is through the use of UIC Class II disposal wells, as discussed earlier in this report.

Status of Emerging Produced Fluids Technologies or Practices

This section evaluates the status of emerging technologies and their potential future roles, based on advantages and limitations of each. Most are chemical techniques that require concentration gradients across a semipermeable membrane, and are presently in small-scale use or experimental research and development phases. An overview of the classes of technologies being researched or under development and a summary of viable advantages and disadvantages of each approach are also presented.

Electrochemical Processes

Electrochemical processes separate dissolved ions from water through ion-permeable membranes or conductive adsorbers through the use of an electrical potential gradient. A summary description of each technology is provided below.[80]

- **Electrodialysis (ED).** An ED unit consists of a series of anion exchange membranes (AEM) and cation exchange membranes (CEM) arranged in an alternating mode between anode and cathode. Positively charged cations migrate toward the cathode, pass the CEM, and are then rejected by the AEM. The opposite occurs when negatively charged anions migrate to the anode. This results in an alternating increasing ion concentration in one compartment (concentrate) and decreasing concentration in the other (diluate).
- **Electrodialysis reversal (EDR).** The EDR process is similar to the ED process, except that it also uses periodic reversal of polarity to minimize membrane scaling and fouling, thus allowing higher water recoveries.

- **Electrodeionization (EDI).** This is an existing commercial desalination technology that combines ED and conventional ion exchange technologies. A mixed-bed ion exchange resin or fiber is placed into the diluate cell of a conventional electrodialysis cell unit to increase the conductivity in the substantially nonconductive water. The process can be performed continuously without chemical regeneration of the ion exchange resin, and can reduce the energy consumption when treating low-salt solutions.
- **Capacitive deionization (CDI).** CDI is an emerging desalination technology where ions are adsorbed onto the surface of porous carbon electrodes (e.g., activated carbon) by applying a low-voltage electric field, thus producing deionized water.

Electrochemical charge-driven separation processes are typically used in desalination of brackish, not highly saline water, significantly reducing the applicability of these technologies to most shale plays. The cost and energy consumption of these processes increase substantially with increasing salinity or TDS concentration. These processes are less prone to fouling as compared to reverse osmosis and nano-filtration membranes. However, low-solubility inorganic salts (e.g., calcium sulfate, calcium carbonate) and multivalent ions (e.g., iron and manganese) can scale the membranes; thus requiring pretreatment.

Ceramic Microfiltration/Ultrafiltration Membrane

Ceramic ultrafiltration and microfiltration membranes consist of a tubular configuration where the feedwater flows inside the membrane channels and permeates through the media to the outside to remove particulates, organic matter, oil and grease, and metal oxides. Due to their extreme stability in harsh environments, ceramic membrane has been reported to be a promising way for produced water purification.[81] Pretreatment using chemical precipitation or a strainer or cartridge filter is necessary as pretreatment for ceramic membranes. Energy requirements for ceramic membranes are lower than those required for polymeric membranes, but ceramic membranes have a higher capital cost than polymeric membranes.[82] The application of ceramic membranes for produced water treatment may increase as more research and pilot studies are conducted.

Membrane Distillation

Membrane distillation (MD) is a thermally driven separation (microfiltration) process, in which only vapor molecules are able to pass through a porous hydrophobic membrane driven by the vapor pressure difference existing between the porous hydrophobic membrane surfaces.[83] MD is the only membrane process that can maintain process performance (i.e., water flux and solute rejection) almost independently of feed solution TDS concentration.[84] MD is capable of producing ultra-pure water at a lower cost compared to conventional distillation processes, and is flexible for most variations in produced feedwater quality and quantity.[85]

Forward Osmosis

Forward osmosis (FO) is a developing membrane process technology that treats wastewater and requires no energy to push the flow through the membrane system, thereby lowering operational costs. A draw solution is employed across the alternate side of the membrane to generate a pressure gradient with a higher pressure on the side containing the waste stream.[86] The membranes used for this process are dense, nonporous barriers similar to reverse osmosis (RO) and nano-filtration (NF) membranes, but are composed of a hydrophilic, cellulose acetate active layer.[87] Typically, the FO draw solution is composed of sodium chloride, but other draw solutions (e.g., ammonium hydrocarbonate, sucrose, and magnesium chloride) have been proposed. During FO, the feed solution is concentrated while the draw solution becomes diluted, and thus must be continuously reconcentrated for sustainable system operation. A challenge is the amount of energy needed to regenerate the draw solution; if waste heat is available the energy inputs to the process can be reduced. One option is the use of RO for reconcentrating the draw solution and producing fresh product water for beneficial use or discharge. FO membranes may be capable of operating with a wide variety of produced fluids with TDS ranging from 500 milligrams per liter to more than 100,000 milligrams per liter, and are capable of rejecting all particulate matter and almost all dissolved constituents (greater than 95% rejection of TDS).[88] These attributes also allow FO to achieve very high theoretical recoveries while minimizing energy and chemical demands; in practice, the recovery rate may be closer to 70%.[89]

Conclusion and Future Considerations

Common approaches for shale energy water management have included trucking of water from the source to the site; storing water in lined, earthen impoundments; and recycling of some portion of produced fluids for reuse in hydraulic fracturing, either at a fixed site or in the field, with the remainder of the fluids disposed of through injection wells or by some other treatment and disposal method. This type of water management approach has limitations, including the production of wastes requiring disposal and the use of significant volumes of fuel for water and waste transport, typically at significant cost. In order to make the process more cost-effective with less environmental impact, new approaches are being sought—for example, the use of lesser-quality sources of water, piping of water where possible, modular water storage, and recycling of produced fluids. Chemical precipitation for scalant removal and mechanical vapor recompression for desalination appear to be the most widely used treatment approaches to date; however, emerging technologies including electrochemical treatment and forward osmosis appear promising. The use of UIC wells for disposal is also heavily relied upon, especially in Ohio and Texas, as a means to manage the portion of produced fluids not being recycled. Long-term viability and capacity of disposal wells are an area of active research to better understand the sustainability of this practice in various shale plays.

Technological progress or changes in water management practices could address some of the most visible impacts on water resources and reduce the risk of impacts on groundwater and surface water quantity and quality. Widespread adoption of fracturing practices that minimize the use of freshwater (groundwater, surface waters, or municipally sourced waters) may reduce pressures from the shale energy sector on scarce water supplies in more arid areas such as Texas and the Rocky Mountain states. In the Appalachian region, overall water supplies are not scarce, but the transportation of water from source to drilling site can involve high trucking costs. Wastewater management practices that minimize the handling of produced fluids and the use of multiple transportation and storage modalities could reduce the risk of impacts to water supplies. Adoption of drilling and completion practices that are less water-intensive and that minimize truck transportation could benefit water quality through reduced erosion along dirt and gravel access roads constructed alongside streams.

While reduction of stresses on water supplies and water quality would represent an environmentally positive step, it is important to realize that water management issues will not disappear entirely. Some freshwater will still be required for shale energy production—for example, in mixtures with flowback water for reuse in subsequent fracturing jobs. The shale energy sector is increasingly recognized as a water consumer (alongside agriculture, municipalities, industry, and electric utilities and other forms of energy production and conversion) in regional water planning and state and local allocation practices.

Adoption of emerging technologies and processes that minimize the water-use intensity of fracturing will have its own challenges, beyond the key issue of cost. For example, public perception is an important consideration in determining which technologies or processes are ultimately adopted and widely deployed. The use of carbon dioxide foam for fracturing, for example, replaces one intensive transportation need with another, since trucks will be required or a dedicated pipeline network will need to be built to deliver the fracking fluid commodity to the drilling location. Industry may be hesitant to adopt alternative technologies and processes if their use reduces energy production or increases costs from shale energy formations.

There continue to be fundamental uncertainties surrounding the acceptable or optimal chemical composition of fracking fluids that would meet emerging environmental concerns but still be effective fluids for hydraulic fracturing and shale oil and gas extraction. For example, even when treated to drinking water standards, acidic mine drainage may still have high sulfate concentrations that increase the potential for downhole precipitation with metals. Metal precipitation could cause plugging of fractures, thereby lowering rates of oil and gas production. The treatments required to lower sulfate concentrations in abandoned mine drainage, and even the extent to which different sulfate concentrations are associated with higher or lower oil and gas production rates, are uncertain and require more study.

The equipment, personnel, and other capital needed for the production of shale energy are highly mobile. Costs can increase or decrease as regional shale development patterns shift. Drilling rigs tend to be moved to those areas with the highest economic returns (for example, away from dry gas to oil producing areas). The mobility of drilling capital suggests that the demand for fracking fluids and wastewater management or treatment services will vary over the course of years or even months. Most treatment facilities, on the other hand, are built in fixed locations, and movement of treatment facilities imposes high costs. Mobile treatment facilities could be developed, but first-generation systems would likely have high costs due to first-of-a-kind engineering and an inability to take advantage of scale economies in water treatment. Variable demand for such facilities may imply that truck transportation, which can be costly and variable, will likely continue to be used until the costs of mobile treatment facilities decline.

Water management issues are relevant to the entire life cycle of shale energy development, because fluids will continue to be produced even after a well is drilled, fractured, and producing oil and/or natural gas. There also are multiple pathways for potential freshwater contamination. Therefore, research that views shale energy production in a life-cycle and materials-flow context may facilitate the identification of technologies and processes that can mitigate potential impacts along different stages of the shale energy development life cycle.

End Notes

[1] Shale oil and gas development also poses challenges to air quality, land management, and other environmental issues, which are not discussed in this report.

[2] In this report, "shale oil" refers to the naturally occurring petroleum extracted from tight shale formations (sometimes called "oil-bearing shales") utilizing hydraulic fracturing methods. Shale oil, as used in this report, is distinct from "oil shale," which refers to unconventional oils extracted from rock formations through pyrolysis.

[3] For the purposes of this report, hydraulic fracturing refers to the process of injecting fluid and a proppant, such as sand, at high pressure into a geologic formation for the purpose of fracturing the rock to allow natural gas or oil to flow from the formation into the production well and be recovered at the surface. (Proppants are particles, such as sand or ceramic beads, that are mixed with fracturing fluid to hold fractures open after a hydraulic fracturing treatment.)

[4] Methane is the main component of natural gas. Methane leakage and migration pathways have received considerable attention, in part because of the difficulty in tracking, monitoring, and attributing specific methane incidents to shale energy activities. Incidents of "stray gas" (attributed to methane migration in the subsurface) in Pennsylvania, for example, have generally not been associated with drilling shale gas wells per se, but with preexisting methane in subsurface waters or from improperly abandoned and unknown older wells. However, some studies have linked increased levels of natural gas in groundwater with proximity to natural gas wells. See Robert B. Jackson et al., "Increased Stray Gas Abundance in a Subset of Drinking Water Wells Near Marcellus Shale Gas Extraction," *Proceedings of the National Academy of Sciences*, June 24, 2013, http://www.pnas.org/content/early/2013/06/19/ 1221635110?tab=author-info.

[5] However, methane is a significantly more potent greenhouse gas than carbon dioxide (although less persistent in the atmosphere); consequently, the advantage of natural gas over other fossil fuels would depend partly on the amount of fugitive methane emissions from the natural gas sector. CRS Report R42986, *An Overview of Air Quality Issues in Natural Gas Systems*, by Richard K. Lattanzio, provides an introduction to emissions associated with natural gas systems. Several studies assess and compare life-cycle emissions of both conventional air pollutants and greenhouse gases (GHGs) from production and use of various fossil fuels. A summary analysis of greenhouse gas analyses can be found in Weber, Christopher, and Christopher Clavin, 2012, "Life Cycle Carbon Footprint of Shale Gas: Review of Evidence and Implications," *Environmental Science and Technology* 46, pp. 5688-5695.

[6] In addition to the Energy Information Administration (EIA) map presented in Figure 1, the U.S. Geological Survey produces a report and map on shale gas assessments; see Laura R.H.

Biewick, compiler, *Map of Assessed Shale Gas in the United States, 2012*, U.S. Geological Survey, Digital Data Series 69-Z, 2013, http://pubs.usgs.gov/dds/dds-069/dds069-z/.

[7] The website FracFocus includes a list and description of the typical chemical additives used for hydraulic fracturing: http://fracfocus.org/chemical-use/what-chemicals

[8] For a discussion of natural gas markets, infrastructure, and related issues, see CRS Report R43636, *U.S. Shale Gas Development: Production, Infrastructure, and Market Issues*.

[9] Forced pooling is also called "compulsory unitization." Such policies are intended to develop resources most efficiently, based on the geology of the resource deposit.

[10] Resources for the Future (RFF), 2012, "A Review of Shale Gas Regulations by State," available at http://www.rff.org/centers/energy last viewed December 22, 2012.

[11] See, for example, New York State Department of Environmental Conservation, *Fact Sheet: What We Learned from Pennsylvania*, NYS DEC NEWS, http://www.dec.ny.gov/energy . Beyond water quality issues, emissions of air pollutants and land-use changes also have generated significant concern for communities and landowners.

[12] Economic and Social Research Council (United Kingdom), 2012, "Fracking and Public Dialogue," available at http://www.esrc.ac.uk/impacts-and-findings/features-casestudies /features/20493/carousel-fracking-and-publicdialogue.aspx, last viewed December 22, 2012.

[13] Osborn, Stephen, Avner Vengosh, Nathaniel Warner, and Robert Jackson, 2011, "Methane Contamination of Drinking Water Accompanying Gas-Well Drilling and Hydraulic Fracturing," *Proceedings of the National Academy of Sciences* vol. 108, no.20, http://www.pnas.org/content/108/20/8172; Nathaniel R. Warner et al., "Geochemical Evidence for Possible Natural Migration of Marcellus Formation Brine to Shallow Aquifers in Pennsylvania," *Proceedings of the Natural Academy of Sciences*, vol. 109, no. 30, July 24, 2012, available at http://www.pnas.org/ content/109/30/11961.

[14] For example, Mark Drajem, "Pennsylvania Fracking Can Put Water at Risk, Duke Study Finds," *Bloomberg Businessweek*, July 10, 2012, http://www.businessweek.com/news/2012-07-09/pennsylvania-fracking-can-put-watersources-at-risk

[15] For example, Rachel Nuwer, "Fracking Did Not Sully Aquifers, Limited Study Finds," *New York Times*, July 9, 2012, http://green.blogs

[16] Robert B. Jackson et al., "Increased Stray Gas Abundance in a Subset of Drinking Water Wells Near Marcellus Shale Gas Extraction," *Proceedings of the National Academy of Sciences*, June 24, 2013, http://www.pnas.org/content/early/ 2013/06/19/1221635110?tab=author-info.

[17] Information on the EPA study is available at http://www2.epa.gov/hfstudy. EPA plans to issue a final report of results by the end of 2016.

[18] Jim Richenderfer, "Water Acquisition for Unconventional Natural Gas Development within the Susquehanna River Basin," Summary of the Technical Workshop on Water Acquisition Modeling: Assessing Impacts through Modeling and Other Means, June 4, 2013, pp. A-16, http://www2.epa.gov/sites/production/files/2013-09/documents/technicalworkshop-water. Based on analysis of SRBC permitting information, approximately 200 surface water intakes allowing in excess of 100 million gallons per day exist. Based on SRBC records, five basin groundwater wells are permitted for shale development. The limited use of groundwater in the Marcellus region is due in part to availability of surface waters and a lack of prolific aquifers co-located with shale resources.

[19] The Barnett play, however, located in central Texas, uses only 20% groundwater (Nicot, J.P., Reedy, R.C., Costley, R.A., and Huang, Y., 2012, *Oil & Gas Water Use in Texas: Update to*

the 2011 Mining Water Use Report, prepared for Texas Oil & Gas Association, Austin, Texas).

[20] Yoxtheimer, D., S. Blumsack, T. Murphy, 2012, "The Decision to Utilize Acidic Coal-Mine Drainage for Hydraulic Fracturing of Unconventional Shale-Gas Wells," *Environmental Practice* 14:4, 7 p.

[21] Cited in Nathan Richardson et al., *The State of State Shale Gas Regulation*, Resources for the Future, May 2013, p. 41.

[22] Colorado Oil and Gas Association, 2011, *Produced Water Fast Facts*; U.S. EPA, 2011, *Proceedings of the Technical Workshops for the Hydraulic Fracturing Study: Water Resources Management.*

[23] Nicot, J.P., Reedy, R.C., Costley, R.A., and Huang, Y., 2012, *Oil & Gas Water Use in Texas: Update to the 2011 Mining Water Use Report*, prepared for Texas Oil & Gas Association, Austin, TX.

[24] Yoxtheimer, D., S. Blumsack, T. Murphy, 2012, "The Decision to Utilize Acidic Coal-Mine Drainage for Hydraulic Fracturing of Unconventional Shale-Gas Wells," *Environmental Practice* 14:4, 7 p.

[25] Under the Comprehensive Environmental Response, Compensation, and Liability Act (CERCLA), any party drawing AMD water from an abandoned mine may be considered an "operator" and subject to the law's liability provisions. Parties could also be in violation of the Clean Water Act for any unpermitted discharges of AMD water to surface waters.

[26] State Impact Pennsylvania, 2013, "Using Abandoned Mine Drainage to Frack," available at http://stateimpact.npr.org/pennsylvania/2013/03/12/using-abandoned-mine-drainage

[27] Ibid.

[28] Kepler, D., and M. Clinger, 2012, *Managing Costs Through Centralization*, presentation given at the Shale Gas Water Management Initiative, Canonsburg, PA, March 2012.

[29] Ibid.

[30] Ibid.

[31] Yeager, 2011.

[32] Nathan Richardson et al., *The State of State Shale Gas Regulation*, Resources for the Future, May 2013, http://www.rff.org/centers/energy

[33] Abdalla, C.W., Drohan, J.R., and Becker, J.C., 2010, *River Basin Approaches to Water Management in the Mid-Atlantic States*, Penn State University Cooperative Extension Publication, 26 p. The Interstate Commission on the Potomac River Basin (ICPRB) has not yet had to deal with shale gas development demands (due in large part to a moratorium on shale gas development in Maryland).

[34] Acharya, H.A., Henderson, C., Matis, H., Kommepalli, H., Moore, B., Wang, H., 2011, *Cost Effective Recovery of Low-TDS Frac Flowback Water for Re-use*, U.S. Department of Energy: DE-FE0000784 Final Report.

[35] Eric Schramm, *What is Flowback, and How Does it Differ from Produced Water?*, Institute for Energy and Environmental Research of Northeastern Pennsylvania Clearinghouse website, March 24, 2011, http://energy

[36] The Safe Drinking Water Act of 1974 (SDWA; P.L. 93-523), as amended, directed EPA to establish an underground injection control (UIC) regulatory program to protect underground sources of drinking water. UIC provisions are contained in SDWA Part C, §§1421-1426; 42 U.S.C. §§300h-300h-5. Class II injection wells discussed here are those wells used for disposal of brines and other wastewater associated with oil and gas production (Class IId).

[37] For example, a 2010 Memorandum of Agreement between the West Virginia Division of Highways and the West Virginia Department of Environmental Protection allows the

beneficial use of gas-well brines within the state for roadway anti-icing and deicing. http://www.dep.wv.gov/WWE/Documents/ WVDOHWVDEP%20Salt%20Brine%20Agreement.pdf. Reuse of brines for road treatment also may pose some runoff or infiltration risks to nearby freshwater bodies or aquifers, which have raised interest in identifying best practices for minimizing such risks.

[38] U.S. Environmental Protection Agency, *Class II Wells—Oil and Gas Related Wells*, http://water groundwater/uic/class2/index.cfm.

[39] Nicot, J.P., Reedy, R.C., Costley, R.A., and Huang, Y., 2012, *Oil & Gas Water Use in Texas: Update to the 2011 Mining Water Use Report*, prepared for Texas Oil & Gas Association, Austin, TX.

[40] See, for example, J. D. Arthur, Stephen L. Dutnell, and David B. Cornue, *Siting and Permitting of Class II Brine Disposal Wells Associated with Development of the Marcellus Shale*, Society of Professional Engineers, SPE 125286, September 2009.

[41] Pennsylvania Department of Environmental Protection, Office of Oil and Gas Management, Production Records accessed October 2012, available at https://www.paoilandgasreporting. state Agreement.aspx.

[42] Pennsylvania Department of Environmental Protection, Office of Oil and Gas Management, Production Records accessed October 2012, available at https://www.paoilandgasreporting. state Agreement.aspx.

[43] U.S. Energy Information Administration, 2012, *Annual Energy Outlook 2012.*

[44] Induced seismicity refers generally to earthquakes that result from human activity. These are typically small tremors that are not felt at the Earth's surface, and which could result from mining activities, filling of large water reservoirs behind dams, and as discussed in this report, from injection of waste fluids from oil and gas activities. Some of the earthquakes from deep well disposal have been large enough to be felt and cause minor damage on ground surface. Typically, microearthquakes caused by the hydraulic fracturing process itself are too small to be felt or cause damage.

[45] Holland, A., 2011, *Examination of possibly induced seismicity from hydraulic fracturing in the Eola Field, Garvin County, Oklahoma*, Oklahoma Geological Survey Open File Report OF1-2011, 28 p.

[46] National Research Council, 2012, *Induced Seismicity Potential in Energy Technologies*, National Academy of Sciences, 300 p. Also, there are nearly 150,000 Class II (brine) injection wells in the United States.

[47] See Zoback, M. L., and M. D. Zoback, 1980, "State of stress in the conterminous United States," *Journal of Geophysical Research*, 85, B11, pp. 6113-6156, http://dx.doi.org/10.1029/JB085iB11p06113; and Zoback, M., 2012, "Managing the seismic risk posed by wastewater disposal," *Earth*, American Geological Institute, http://www.earthmagazine.org/article/managing-seismic-risk-posed-by-wastewater-disposal.

[48] Frohlich, C., 2012, "A survey of earthquakes and injection well locations in the Barnett Shale, Texas," *Leading Edge*, 1446-1451.

[49] For an overview of the scientific understanding of induced seismicity in the United States as of July 2013, see William L. Ellsworth, "Injection-induced Earthquakes," *Science*, vol. 341 (July 12, 2013).

[50] See Frohlich, C., Hayward, C., Stump, B., and Potter, E., 2011, "The Dallas-Fort Worth Earthquake Sequence: October 2008 through May 2009," *Bulletin of the Seismological Society of America*, 101(1), pp. 327-340; Frohlich, C., 2012, "A survey of earthquakes and injection well locations in the Barnett Shale, Texas," *Leading Edge*, pp. 1446- 1451; and

Shemeta, J.E., Eide, E.A., Hitzman, M.W., Clarke, D.D., Detournay, E., Dieterich, J.H., Dillon, D.K. Green, S.J., Habiger, R.M., McGuire, R.K., Mitchell, J.K., Smith J.L., Ortego, J.R., and Gibbs, C.R., 2012, "The potential for induced seismicity in energy technologies," *Leading Edge*, Society of Exploration Geophysicists, v. 31, no. 12, pp. 1438-1443.

[51] Ohio Division of Natural Resources, 2012, *Preliminary Report on the Northstar Class II Injection Well and the Seismic Events in the Youngstown, Ohio, Area*, 24 p.

[52] *State Oil and Gas Regulations Designed to Protect Water Resources*, 2009, U.S. Department of Energy National Energy Technology Laboratory and Groundwater Protection Council, 62 p.

[53] Pennsylvania Department of Environmental Protection (PA DEP), 2009, *Stray Natural Gas Migration Associated with Oil and Gas Wells*; PA DEP: Harrisburg, 2009, http://www.dep.state Stray%20Gas%20Migration%20Cases.pdf; Pittsburgh Geological Society, *Natural Gas Migration Problems in Western Pennsylvania*.

[54] http://stateimpact.npr.org/pennsylvania/2012/10/10/perilous-pathways-

[55] Closed loop, or reduced emission completions (RECs), is a term used to describe a practice that captures air pollutants and gas produced during well completions or well workovers following fracturing. In 2012, EPA issued regulations under the Clean Air Act that require the use of RECs on hydraulically fractured natural gas wells beginning in 2015. See U.S. EPA, *Reduced Emissions Completions for Hydraulically Fractured Natural Gas Wells*, http://www.epa.gov/gasstar/documents/reduced_emissions_completions.pdf.

[56] Performance is an important issue, and experimentation is costly when unconventional wells cost more than $4 million to $6 million. See Fisher, K., 2012, "Green frac fluid chemistry optimizes well productivity, environmental performance," *American Oil and Gas Reporter*, http://www.aogr.com, March 2012.

[57] A comparison of regulations by state as of July 2012 can be found in Resources for the Future, 2012, *Managing the Risks of Shale Gas Development: Identifying a pathway toward responsible development: A review of shale gas regulations by state*.

[58] For example, Halliburton has developed "Clean Stim" as a fluid using only "food industry" ingredients. Information is available at http://www.halliburton.com/en-US/ps/stimulation

[59] FracFocus is an online repository (available at http://www.fracfocus.com) for disclosure of chemical constituents in hydraulic fracturing fluids, and is operated by the Groundwater Protection Council and the Interstate Oil and Gas Compact Commission.

[60] A list of all chemicals used in hydraulic fracturing from 2005 through 2011 can be found in Appendix A of the EPA "Study of the Potential Impacts of Hydraulic Fracturing on Drinking Water Resources: Progress Report," December 2012, available at http://www.epa.gov/hfstudy. EPA has not yet made a determination of toxicity for these additives.

[61] As of the end of 2013, the most recent set of draft rules is available at http://www.blm.gov/pgdata/etc/medialib/blm/wo/Communications_Directorate/public_affairs/hydraulicfracturing.Par.91723.File.tmp/HydFrac_SupProposal.pdf.

[62] Guidance from the American Petroleum Institute (API) suggests that operators should disclose information on chemical additives and their composition when requested, and that "best practice is to use additives that pose minimal risk of possible adverse human health effects to the extent possible in delivering needed fracture effectiveness."

[63] See Fisher, K., 2012, "Green frac fluid chemistry optimizes well productivity, environmental performance," *American Oil and Gas Reporter*, http://www.aogr.com, March 2012.

[64] Joint Institute for Strategic Energy Analysis, 2012, *Natural Gas and the Transformation of the U.S. Electricity Sector*, National Renewable Energy Laboratory Report NREL/TP-6A50-55538, available at http://www.nrel.gov/docs/ fy13osti/55538.pdf.

[65] Sorbtion is a process by which one substance becomes attached to another and includes adsorption (adherence onto the surface of another substance) and absorption (incorporation into a substance of a different state (e.g., a liquid being absorbed by a solid).

[66] Tight sand gas accumulations occur where gas migrates from a source rock into a sandstone formation with relatively low permeability compared to other "conventional" sandstone formation reservoirs. The relatively low permeability of the tight sandstone limits the ability of the gas to migrate further upward without an enhanced recovery technique such as fracturing.

[67] Joshi, S.D., 2007, "Reservoir aspects of horizontal and multilateral wells," 265 p., SPE Ann. Tech. convention, 2007; Greenberg, J., 2012, "Today's technologies support operator goals," 2012 *North American Unconventional Yearbook: Technology*, http://www.hart energy.

[68] Devon's green-completion process, which involves capturing methane rather than flaring, is described at http://www.dvn.com/CorpResp/initiatives/Pages/GreenCompletions.aspx.

[69] U.S. Energy Information Administration, *Today in Energy*, republished March 21, 2014, http://www.eia.gov/ todayinenergy/detail.cfm?id=15511&src.

[70] U.S. EPA, "Oil and Natural Gas Sector: New Source Performance Standards and National Emission Standards for Hazardous Air Pollutants Reviews, Final Rule," 77 *Federal Register* 49489, August 16, 2012.

[71] For more analysis, see CRS Report R42833, *Air Quality Issues in Natural Gas Systems*, by Richard K. Lattanzio.

[72] Colorado Oil and Gas Association, 2011. *Produced Water Fast Facts*; U.S. EPA, 2011, *Proceedings of the Technical Workshops for the Hydraulic Fracturing Study: Water Resources Management.*

[73] Nicot, J.P., Reedy, R.C., Costley, R.A., and Huang, Y., 2012, *Oil & Gas Water Use in Texas: Update to the 2011 Mining Water Use Report*, prepared for Texas Oil & Gas Association, Austin, TX.

[74] U.S. Department of Energy National Energy Technology Laboratory Project DE-FE0001466, 2012.

[75] Emamjomeh, M., Sivakumar, M., 2009, "Review of pollutants removed by electrocoagulation and electrocoagulation/flotation processes," *Journal of Environmental Management* 90, pp. 1663-1679.

[76] Acharya, H.A., Henderson, C., Matis, H., Kommepalli, H., Moore, B., Wang, H., 2011, *Cost Effective Recovery of Low-TDS Frac Flowback Water for Re-use*, Department of Energy: DE-FE0000784 Final Report.

[77] U.S. Department of Energy National Energy Technology Laboratory, RPSEA Project 07122-12, 2009, *An Integrated Framework for Treatment and Management of Produced Water.*

[78] U.S. Department of Energy National Energy Technology Laboratory, RPSEA Project 08122-36, 2012, *Produced Water Pretreatment for Water Recovery and Salt Production.*

[79] U.S. Department of Energy National Energy Technology Laboratory, RPSEA Project 07122-12, 2009, *An Integrated Framework for Treatment and Management of Produced Water.*

[80] Much of the technical description in this section is based upon Drewes, J., T.Y. Cath, P. Xu, J. Graydon, J. Veil, S. Snyder, 2008, *An Integrated Framework for Treatment and Management of Produced Water*, RPSEA Forum, Golden, CO, available at http://www.rpsea.org/attachments/wysiwyg/681/cath1.pdf.

[81] Zhang, H., Zhong, Z., Xing, W., 2013, "Application of ceramic membranes in the treatment of oilfield-produced water: Effects of polyacrylamide and inorganic salts," *Desalination* 309, pp. 84-90.

[82] Drewes, J., T.Y. Cath, P. Xu, J. Graydon, J. Veil, S. Snyder, 2008, *An Integrated Framework for Treatment and Management of Produced Water*, RPSEA Forum, Golden CO, available at http://www.rpsea.org/attachments/wysiwyg/ 681/cath1.pdf.

[83] Alkhudhiri, A., Darwish, N., Hilal, N., 2013, "Produced water treatment: Application of Air Gap Membrane Distillation," *Desalination* 309, pp. 46-51.

[84] Drewes, J., T.Y. Cath, P. Xu, J. Graydon, J. Veil, S. Snyder, 2008, *An Integrated Framework for Treatment and Management of Produced Water*, RPSEA Forum, Golden CO, available at http://www.rpsea.org/attachments/wysiwyg/ 681/cath1.pdf.

[85] Alkhudhiri, A., Darwish, N., Hilal, N., 2013, "Produced water treatment: Application of Air Gap Membrane Distillation," *Desalination* 309, pp. 46-51.

[86] Olawoyin, R., Madu, C., Enab, K., 2012, "Optimal Well Design for Enhanced Stimulation Fluids Recovery and Flow-back Treatment in the Marcellus Shale Gas Development using Integrated Technologies," *Hydrology Current Research* 3:141. doi:10.4172/2157-7587.1000141.

[87] Drewes, J., T.Y. Cath, P. Xu, J. Graydon, J. Veil, S. Snyder, 2008, "An Integrated Framework for Treatment and Management of Produced Water," RPSEA Forum, Golden CO, available at http://www.rpsea.org/attachments/ wysiwyg/681/cath1.pdf.

[88] Olawoyin, R., Madu, C., Enab, K., 2012, "Optimal Well Design for Enhanced Stimulation Fluids Recovery and Flow-back Treatment in the Marcellus Shale Gas Development using Integrated Technologies," *Hydrology Current Research* 3:141. doi:10.4172/2157-7587.1000141. Bryan D. Coday et al., "The sweet spot of forward osmosis: Treatment of produced water, drilling wastewater, and other complex and difficult liquid streams," *Desalination*, no. 333 (2013), pp. 23-35.

[89] Bryan D. Coday et al., "The sweet spot of forward osmosis: Treatment of produced water, drilling wastewater, and other complex and difficult liquid streams," *Desalination*, no. 333 (2013), pp. 23-35.

In: Unconventional Oil and Shale Gas
Editor: Amber L. Tuft

ISBN: 978-1-63482-130-8

Chapter 3

HYDRAULIC FRACTURING AND SAFE DRINKING WATER ACT REGULATORY ISSUES*

Mary Tiemann and Adam Vann

SUMMARY

Hydraulic fracturing is a technique developed initially to stimulate oil production from wells in declining oil reservoirs. With technological advances, hydraulic fracturing is now widely used to initiate oil and gas production in unconventional (low-permeability) oil and gas formations that were previously uneconomical to produce. This process now is used in more than 90% of new oil and gas wells and in many existing wells to stimulate production. Hydraulic fracturing is done after a well is drilled, and involves injecting large volumes of water, sand (or other propping agent), and specialized chemicals under enough pressure to fracture the formations holding the oil or gas. The sand or other proppant holds the fractures open to allow the oil or gas to flow freely out of the formation and into a production well. Its application, in combination with horizontal drilling, for production of natural gas (methane) from tight gas sands, unconventional shale formations, and coal beds, has resulted in the marked expansion of estimated U.S. natural gas reserves in recent years. Similarly, hydraulic fracturing is enabling the development of tight oil

* This is an edited, reformatted and augmented version of a Congressional Research Service publication, No. R41760, dated June 17, 2014.

resources, such as the Bakken and Eagle Ford formations. The rapid growth in the use of fracturing has raised concerns over its potential impacts on groundwater and drinking water sources, and has led to calls for more state and/or federal oversight of this activity.

Historically, the Environmental Protection Agency (EPA) had not regulated the underground injection of fluids for hydraulic fracturing of oil or gas production wells. In 1997, the U.S. Court of Appeals for the 11th Circuit ruled that fracturing for coalbed methane (CBM) production in Alabama constituted underground injection and must be regulated under the Safe Drinking Water Act (SDWA). This ruling led EPA to study the risk that hydraulic fracturing for CBM production might pose to drinking water sources. In 2004, EPA reported that the risk was small, except where diesel was used, and that national regulation was not needed. However, to address regulatory uncertainty the ruling created, the Energy Policy Act of 2005 (EPAct 2005) revised the SDWA term "underground injection" to explicitly exclude the injection of fluids and propping agents (except diesel fuels) used for hydraulic fracturing purposes. Thus EPA lacks authority under the SDWA to regulate hydraulic fracturing, except where diesel fuels are used. In February 2014, EPA issued final permitting guidance for hydraulic fracturing operations using diesel fuels.

As the use of the process has grown, some in Congress would like to revisit the 2005 statutory exclusion. Legislation to revise the act's definition of underground injection to explicitly include hydraulic fracturing has been offered in recent years, but not enacted. A variety of hydraulic fracturing bills are pending in the 113th Congress. In EPA's FY2010 appropriations act, Congress urged the agency to study the relationship between hydraulic fracturing and drinking water quality. In 2012, EPA issued a research progress report. The agency expects to issue a final report in 2016.

This report reviews past and proposed treatment of hydraulic fracturing under the SDWA, the principal federal statute for regulating the underground injection of fluids to protect groundwater sources of drinking water. It reviews current SDWA provisions for regulating underground injection activities, and discusses some possible implications of the enactment of legislation authorizing EPA to regulate hydraulic fracturing (beyond diesel) under this statute. The report also reviews legislative proposals concerning the regulation of hydraulic fracturing under the SDWA.

INTRODUCTION

Hydraulic Fracturing in Oil and Gas Production

The process of hydraulic fracturing was developed initially in the 1940s to stimulate production from oil reservoirs with declining productivity.[1] More recent technological advances in hydraulic fracturing, along with horizontal drilling, have allowed this practice to be used to initiate oil and gas production in unconventional (low-permeability) oil and gas formations.[2] Its application in the production of natural gas from coal beds, tight gas sands,[3] and unconventional shale formations has resulted in the marked expansion of estimated U.S. natural gas reserves and production in recent years. Similarly, hydraulic fracturing has enabled the development of domestic tight oil resources, such as the Bakken Formation in North Dakota and Montana, and the Eagle Ford Formation in Texas. However, the rapidly increasing and geographically expanding use of this well stimulation process has raised concerns over its potential impacts on groundwater and drinking water and has led to calls for greater state and/or federal oversight of hydraulic fracturing and more research on its potential risks to water resources.

Hydraulic fracturing involves injecting into production wells large volumes of water, sand or other proppant,[4] and specialized chemicals under enough pressure to fracture low-permeability geologic formations containing oil and/or natural gas.[5] The sand or other proppant holds the new fractures open to allow the oil or gas to flow freely out of the formation and into a production well. Fracturing fluid and water remaining in the fracture zone can inhibit oil and gas production, and must be pumped back to the surface. The fracturing fluid—"flowback"—along with any naturally occurring formation water pumped to the surface, together called produced water, typically has been disposed of through deep well injection, and less frequently has been treated and discharged into surface waters.[6] According to industry estimates for various geographic areas, the volume of flowback water can range from less than 30% to more than 70% of the original fracture fluid volume.[7] Increasingly, efforts are being made to treat and reuse flowback.

Reliance on the use of hydraulic fracturing continues to increase, as more easily accessible oil and gas reservoirs have declined and companies move to develop unconventional oil and gas formations. The Interstate Oil and Gas Compact Commission (IOGCC) reported that 90% of oil and gas wells in the United States have undergone hydraulic fracturing to stimulate production.[8] According to the American Petroleum Institute (API), hydraulic fracturing has been applied to more than 1 million wells nationwide, and typically multiple times per well.[9] The U.S. Energy Information Administration (EIA) reports that natural gas from tight sand formations was, until recently, the largest source of unconventional production, but has been surpassed by production from shale formations.[10] **Figure 1** illustrates different types of natural gas reservoirs.

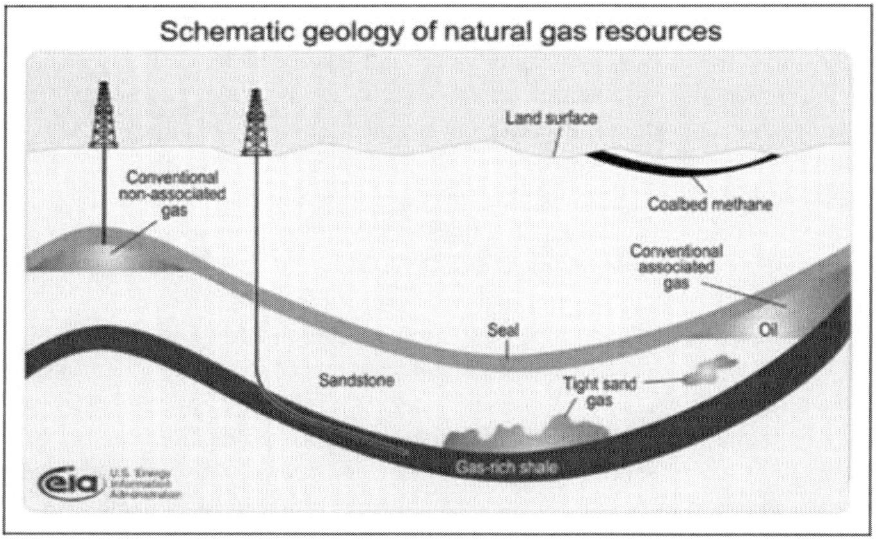

Source: U.S. Energy Information Administration, Independent Statistics and Analysis, October 2008. Available at http://www.eia.gov/oil_gas/natural_gas/special/ngresources/ngresources.html.

Notes: The diagram shows schematically the geologic nature of most major U.S. sources of natural gas:

- Gas-rich shale is the source rock for many natural gas resources, but, until [recently], has not been a focus for production. Horizontal drilling and hydraulic fracturing have made shale gas an economically viable alternative to conventional gas resources.
- Conventional gas accumulations occur when gas migrates from gas rich shale into an overlying sandstone formation, and then becomes trapped by an overlying impermeable formation, called the seal. Associated gas accumulates in conjunction with oil, while non-associated gas does not accumulate with oil.
- Tight sand gas accumulations occur in a variety of geologic settings where gas migrates from a source rock into a sandstone formation, but is limited in its ability to migrate upward due to reduced permeability in the sandstone.
- Coalbed methane does not migrate from shale, but is generated during the transformation of organic material to coal.

Figure 1. Geologic Nature of Major Sources of Natural Gas in the United States.

Production of shale gas and shale oil (often called "tight" oil) involves drilling a well vertically and then drilling horizontally out from the wellbore. Because of the low permeability of these formations, more wells must be drilled into a reservoir than into more permeable, conventional reservoirs to retrieve the same amount of oil or gas. A benefit of horizontal drilling through a producing shale layer is that one well pad that utilizes horizontal well drilling can replace numerous individual well pads and reduce the surface density of wells in an area. Six to eight horizontal wells, and potentially more, can be drilled from a single well pad and access the same reservoir. According to a report prepared for DOE,

> The spacing interval for *vertical* wells in the gas shale plays averages 40 acres per well for initial development. The spacing interval for *horizontal* wells is likely to be approximately 160 acres per well. Therefore, a 640-acre section of land could be developed with a total of 16 vertical wells, each on its own individual well pad, or by as few as 4 horizontal wells all drilled from a single multi-well drilling pad.[11]

A single production well may be fractured multiple times, using from 500,000 gallons to more than 10 million gallons of water, with compounds and proppants of various amounts added to the water. Slickwater fracturing, which involves adding conditioning chemicals to water to increase fluid flow, is a more recent development that has improved production of unconventional shale gas.[12]

Hydraulic Fracturing and Drinking Water Issues

While the use of high-volume hydraulic fracturing has enabled the oil and gas industry to markedly increase domestic production, questions have emerged regarding the potential impacts this process may have on groundwater quality, and particularly on private wells and drinking water supplies. During hydraulic fracturing, new fractures are induced into a shale or other tight formation and existing fractures may be lengthened. As production activities have increased and expanded into more populated areas, so has concern that the fracturing process might introduce chemicals, methane, and other contaminants into aquifers.

A particularly contentious issue concerns whether the fracturing process could create or extend fractures linking the producing zone to an overlying aquifer, and thus provide a pathway for gas or fracturing fluids to migrate. In shale formations, the vertical distance separating the target zone from usable aquifers generally is much greater than the length of the fractures induced during hydraulic fracturing. Thousands of feet of rock layers typically overlay the produced portion of the shale, and these layers serve as barriers to flow. In these circumstances, geologists and state regulators generally view as remote the possibility of creating a fracture that could reach a potable aquifer. If the shallow portions of shale formations were developed, then the thickness of the overlying rocks would be less and the distance from the shale to potable aquifers would be shorter, posing more of a risk to groundwater. In contrast to shale, coalbed methane (CBM) basins often qualify as underground sources of drinking water. Injection of fracturing fluids directly into or adjacent to such formations would be more likely to present a risk of contamination, and this is where initial regulatory attention and study was focused.[13] (See discussion under "EPA's 2004 Review of Hydraulic Fracturing for CBM Production.")

Complaints of impacts to well water have emerged with unconventional gas development and the use of hydraulic fracturing; however, state investigations have not reported a direct connection between hydraulic fracturing of shale formations and groundwater contamination. In 2009, the Ground Water Protection Council (GWPC)[14] reported that several citizen complaints of well water contamination attributed to hydraulic fracturing appeared to be related to hydraulic fracturing of CBM zones that were in relatively close proximity to underground sources of drinking water, although the Environmental Protection Agency's (EPA's) 2004 study found no confirmed cases of contamination.[15]

Regulators have expressed more concern about the potential groundwater contamination risk that is associated with developing a natural gas or oil well (drilling through an overlying aquifer, and casing, cementing, and completing the well). The challenges of sealing off the groundwater and isolating it from possible contamination are common to the development of any oil or gas well, and are not unique to hydraulic fracturing. However, some states have revised cementing and other well construction requirements specifically to address hydraulic fracturing. Also, industry best practices for well construction and integrity have been developed for hydraulic fracturing.[16]

Another potential source of groundwater contamination comes from surface activities. Leaky surface impoundments, accidental spills of hydraulic fracturing fluids, or mismanagement of drilling fluids at the production site all could increase the risk of contamination. Additionally, inadequate wastewater management practices (including the storage, treatment, and disposal of flowback and produced water) can present risks to groundwater.[17]

Identifying the source or cause of groundwater contamination can be difficult for various reasons, including the complexity of hydrogeologic processes and investigations, a lack of baseline testing of nearby water wells prior to drilling and fracturing, and the confidential business information status traditionally provided for fracturing compounds.[18] In cases that have been investigated, regulators typically have determined that groundwater contamination was caused by failure of well-bore casing and cementing, well operation problems, or surface activities, rather than the hydraulic fracturing process. In Pennsylvania, for example, regulators confirmed that methane had migrated to water wells from drilling sites in two counties, and determined that the gas migration was caused by improperly cased and cemented wells and, in some cases, by excessive pressures.[19]

Although regulators have not identified hydraulic fracturing of shale formations as the direct cause of groundwater contamination, water quality problems attributed to other exploration and production activities have raised concerns regarding the adequacy and/or enforcement of state well construction and wastewater management regulations for purposes of managing oil and gas development that is increasingly dependent on high-volume hydraulic fracturing. In the past several years, major producing states have been revising their oil and gas laws and regulations to address hydraulic fracturing more explicitly or comprehensively, and some states have increased the number of inspectors to oversee increased exploration and production activities.[20]

The debate over the groundwater contamination risks associated with hydraulic fracturing has been fueled in part by the lack of scientific studies to assess the practice and related complaints, and in 2009, Congress urged EPA to conduct a study on the relationship between hydraulic fracturing and drinking water.[21] The "hydraulic fracturing" debate also has been complicated by terminology. Many who express concern over the potential environmental issues associated with hydraulic fracturing do not differentiate the well stimulation process of "fracing" from the broader range of activities associated with unconventional oil and gas exploration and production.[22]

Some have called for broader federal regulation of hydraulic fracturing through the Safe Drinking Water Act (SDWA),[23] and legislation has been offered in the past two Congresses to give EPA this authority. Such proposals have prompted debate over the possibility of broad new federal involvement in regulating oil and gas development—an area long managed by the states. In addition to a lack of consensus regarding the federal role, basic regulatory issues contribute to uncertainty over a possible regulatory framework that might be developed for hydraulic fracturing activities under the SDWA. At issue is whether the further federal regulation is needed, and if so, does the current EPA underground injection control (UIC) program under the SDWA fit? EPA developed this program primarily to regulate wells that received fluids injected for the long term or for enhanced recovery operations, but excluded oil and gas *production* wells. This distinction could raise regulatory challenges and the possibility that the agency may need to develop an essentially new framework to address hydraulic fracturing of production wells. In February 2014, EPA issued final guidance for fracturing operations that involve diesel fuels.[24] This guidance may indicate how the agency might approach the broader regulation of hydraulic fracturing if so directed by Congress. (See discussion under "EPA Guidance for Permitting Hydraulic Fracturing Using Diesel Fuels.")

This report reviews past and proposed treatment of hydraulic fracturing under the SDWA, the principal federal statute for regulating the underground injection of fluids to protect groundwater sources of drinking water. It reviews current SDWA provisions for regulating underground injection activities and discusses some possible implications of, and issues associated with, enactment of legislation authorizing EPA to regulate hydraulic fracturing under this statute. This report also discusses recent developments among the states to address the growing reliance on high-volume hydraulic fracturing, which may add insight to the possible implications of proposed federal legislation and any subsequent regulations.

THE SAFE DRINKING WATER ACT (SDWA) AND THE FEDERAL ROLE IN REGULATION OF UNDERGROUND INJECTION

Review of Relevant SDWA Underground Injection Control (UIC) Provisions

To evaluate studies and any new federal action to regulate hydraulic fracturing under the SDWA, it may be helpful to understand the existing statutory and regulatory framework.

Most public water systems and nearly all rural residents rely on groundwater as a source of drinking water. Because of the nationwide importance of underground sources of drinking water, Congress included groundwater protection provisions in the 1974 Safe Drinking Water Act. The SDWA, among other things, directs EPA to regulate the underground injection of fluids (including solids, liquids, and gases) to protect underground sources of drinking water.[25]

Part C of the SDWA establishes the national regulatory program for the protection of underground sources of drinking water, including the oversight and limitation of underground injections that could affect aquifers through the establishment of underground injection control regulations. Key UIC requirements and exceptions contained in SDWA, Part C, include the following:

- Section 1421 of the SDWA directs the EPA Administrator to promulgate regulations for state UIC programs, and mandates that the EPA regulations "contain minimum requirements for programs to prevent underground injection that endangers drinking water sources." Section 1421(b)(2) specifies that EPA may not prescribe requirements for state UIC programs which interfere with or impede—(A) the underground injection of brine or other fluids which are brought to the surface in connection with oil or natural gas production or natural gas storage operations, or (B) any underground injection for the secondary or tertiary recovery of oil or natural gas, *unless such requirements are essential to assure that underground sources of drinking water will not be endangered by such injection.*[26] [Emphasis added.]
- Section 1421(d), as amended by Energy Policy Act of 2005 (EPAct 2005),[27] specifies that the term "underground injection" as it is used in the SDWA means the subsurface emplacement of fluids by well injection, and specifically excludes the underground injection of fluids or propping agents associated with hydraulic fracturing operations related to oil, gas, or geothermal production activities.[28] The use of diesel fuels in hydraulic fracturing, however, forfeits eligibility for this exclusion from the definition of "underground injection."[29]
- Section 1422 authorizes EPA to delegate primary enforcement authority (primacy) for UIC programs to the states, provided that the state program meets EPA requirements promulgated under Section 1421 and prohibits any underground injection that is not authorized by a state permit or rule.[30] If a state's UIC program plan is not approved, or the state has chosen not to assume program responsibility, then EPA must implement the UIC program in that state.

- Section 1425 authorizes EPA to approve the portion of a state's UIC program that relates to "any underground injection for the secondary or tertiary recovery of oil or natural gas" if the state program meets certain requirements of Section 1421 and represents an effective program to prevent underground injection which endangers drinking water sources.[31] Under this provision, states may demonstrate to EPA that their existing programs for oil and gas injection wells are effective in preventing endangerment of underground sources of drinking water. This provides states with an alternative to meeting the specific requirements contained in EPA regulations promulgated under Section 1421.[32] (See discussion on p. 11.)
- Section 1423 authorizes EPA enforcement actions for UIC regulatory violations.
- Section 1431 applies broadly to the SDWA and grants the EPA Administrator emergency powers to issue orders and commence civil actions to protect public water systems or underground sources of drinking water.[33]
- Section 1449, another broadly applicable SDWA provision, authorizes citizen civil actions against persons allegedly in violation of the act's enforceable requirements, or against EPA for allegedly failing to perform a duty. State-administered oil and gas programs may not have such provisions, so this could represent an expansion in the ability of citizens to challenge administration of statutes and regulations related to hydraulic fracturing and drinking water, were the hydraulic fracturing exemption provision to be repealed.

The "Endangerment" Standard

As noted, the SDWA states that UIC regulations must "contain minimum requirements for effective programs to prevent underground injection which endangers drinking water sources."[34] Known as the "endangerment standard," this statutory standard is a major driving force in EPA regulation of underground injection.

The endangerment language focuses on protecting groundwater that is used or may be used to supply public water systems. This focus parallels the general scope of the statute, which addresses the quality of water provided by public water systems and does not address private, residential wells. The endangerment language has raised questions as to whether EPA regulations can reach underground injection activities to protect groundwater that is not used by public water systems.

Defining "Underground Source of Drinking Water"

The SDWA directs EPA to protect against endangerment of an "underground source of drinking water" (USDW). The statute defines a USDW to mean an aquifer or part of an aquifer that either

- supplies a public water system, or
- contains a sufficient quantity of groundwater to supply a public water system;[35] and
    - currently supplies drinking water for human consumption; or
    - contains fewer than 10,000 milligrams per liter (mg/L or parts per million) total dissolved solids; and
- is not an "exempted aquifer."[36]

In a 2004 report on hydraulic fracturing of coalbed methane reservoirs, the agency further noted that the "EPA also assumes that all aquifers contain sufficient quantity of groundwater to supply a public water system, unless proven otherwise through empirical data."[37] However, because these expanded agency characterizations of what constitutes a USDW are not included in SDWA or related regulation, and, therefore, are not binding on the agency, it is uncertain how they might be applied in future situations. Notably, the SDWA does not prohibit states from establishing requirements that are stricter than federal requirements, and many states have their own definitions and classifications for groundwater resources.

UIC Regulatory Program Overview

The UIC program regulates more than 800,000 injection wells. To implement the UIC program as mandated by the provisions of the SDWA described above, EPA has established six classes of underground injection wells based on categories of materials that are injected into the ground by each class. In addition to the similarity of fluids injected in each class of wells, each class shares similar construction, injection depth, design, and operating techniques. The wells within a class are required to meet a set of appropriate performance criteria for protecting underground sources of drinking water (USDW). The six well categories are briefly described below, including the estimated number of wells nationwide.[38]

- Class I wells inject hazardous wastes, industrial non-hazardous liquids, or municipal wastewater beneath the lowermost USDW. (There are 680 such wells regulated as Class I wells in the United States.) The most stringent UIC regulations apply to these wells.
- Class II wells inject brines and other fluids associated with oil and gas production, and hydrocarbons for storage. The wells inject fluids beneath the lowermost USDW (172,068 wells). Section 1425, which allows states to apply their own regulations in lieu of EPA regulations, applies to Class II wells.[39]
- Class III wells inject fluids associated with solution mining of minerals (e.g., salt and uranium) beneath the lowermost USDW (22,131 wells).
- Class IV wells inject hazardous or radioactive wastes into or above USDWs. These wells are banned unless authorized under a federal or state groundwater remediation project (33 wells).
- Class V includes all injection wells not included in Classes I-IV, including experimental wells. Class V wells frequently inject non-hazardous fluids into or above USDWs and are typically shallow, on-site disposal systems. However, some deep Class V wells inject below USDWs (400,000-650,000 wells).[40]
- Class VI wells: In 2010, EPA issued a rule establishing Class VI wells, to be used for the geologic sequestration of carbon dioxide (no permitted wells).

The UIC regulatory program includes the following broad elements: site characterization, area of review, well construction, well operation, site monitoring, well plugging and post-injection site care, public participation, and financial responsibility. While the six classes broadly share similar regulatory requirements, those for Class I wells are the most comprehensive and stringent. **Table 1** outlines the shared minimum technical requirements for Class I, II, and III wells. **Table 2** outlines basic regulatory requirements for Class II wells.

Table 1. Minimum Federal Technical Requirements for Class I, II, and III Wells

Permitting Requirements Common to Class I, II, and III Wells
Demonstration that casing and cementing are adequate to prevent movement of fluid into or between USDWs.
Cement bond logs are often needed to evaluate/verify the adequacy of the cementing records.
Financial assurances (bond, letter of credit, or other adequate assurance) that the owner or operator will maintain financial responsibility to properly plug and abandon the wells.
A maximum operating pressure calculated to avoid initiating and/or propagating fractures that would allow fluid movement into a USDW.
Monitoring and reporting requirements.
Requirement that all permitted (and rule authorized) wells which fail mechanical integrity be shut in immediately. A well may not resume injection until mechanical integrity has been demonstrated.
Schedule for demonstrating mechanical integrity (at least every five years for Class I nonhazardous, Class II, and Class III salt recovery wells).a
All permitted injection wells, which have had the tubing disturbed, must have a pressure test to demonstrate mechanical integrity.
Plans for plugging and abandonment. All Class I, II, and III wells must be plugged with cement.

Source: U.S. Environmental Protection Agency, Technical Program Overview: Underground Injection Control Regulations, Revised 2001, EPA 816-R-02-025, December 2002, p. 65.

a. Class I hazardous wells must demonstrate mechanical integrity once a year.

Table 2. Minimum EPA Regulatory Requirements for Class II Wells

Requirement	Explanation
Permit Required	Yes, except for existing Enhanced Oil Recovery (EOR) wells authorized by rule
Life of Permit	Specific period, may be for life of well
Area of Review	New wells—¼ mile fixed radius or radius of endangerment
Mechanical Integrity Test (MIT)	Internal MIT: prior to operation, and pressure test or alternative at least

Table 2. (Continued)

Requirement	Explanation
Required	once every five years for internal well integrity. External MIT: cement records may be used in lieu of logs.
Other Tests	Annual fluid chemistry and other tests as needed/required by permit
Monitoring	Injection pressure, flow rate and cumulative volume, observed weekly for disposal and monthly for enhanced recovery
Reporting	Annual

Source: U.S. Environmental Protection Agency, Technical Program Overview: Underground Injection Control Regulations, Revised 2001, EPA 816-R-02-025, December 2002, pp. 11, 67, and Appendix E.

Class II Wells

Because this discussion of hydraulic fracturing is related to oil and gas production, this report focuses primarily on regulatory requirements for Class II wells rather than other categories of injection wells in EPA's UIC program. If authorized or mandated to regulate hydraulic fracturing broadly under SDWA, EPA might regulate hydraulic fracturing as a Class II activity, which would parallel its proposed approach for regulating the injection of diesel for fracturing purposes.[41] However, it is possible that EPA could classify oil and gas production wells that are hydraulically fractured under a different class, or develop an entirely new regulatory structure or subclass of wells.[42]

Class II wells may be used to dispose of brines (salt water) and other fluids associated with oil and gas production or storage, to store natural gas, or to inject fluids for enhanced oil and gas recovery. ER wells inject brine, water, steam, polymers, or carbon dioxide primarily into oil-bearing formations (also called secondary or tertiary recovery). Enhanced recovery injection wells are separate from, and typically surrounded by, production wells.[43]

EPA estimates that approximately 80% of Class II wells are enhanced recovery (ER) wells. For example, Pennsylvania has roughly 1,850 Class II wells—almost all are ER wells and only seven are wastewater disposal wells.) **Figure 2** illustrates the various types of Class II wells.

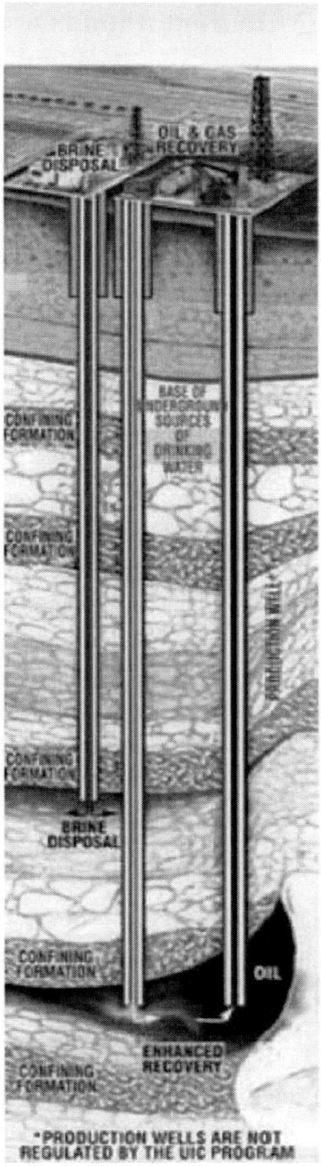

Source: U.S. Environmental Protection Agency.

Note: An oil or gas production well would require a Class II UIC permit if the hydraulic fracturing fluid to be used contains diesel fuels.

Figure 2. Class II Wells.

State Primacy for UIC Program Administration

SDWA Section 1422 authorizes states to assume primary enforcement authority for the UIC program for any or all classes of injection wells. EPA must delegate this authority, provided that the state program meets EPA requirements promulgated under Section 1421 and prohibits underground injection that is not authorized by permit or rule. Otherwise, EPA must implement the UIC program in that state. Thirty-three states have assumed primacy for the entire UIC program (injection well Classes I through V), EPA has lead implementation authority in 10 states, and authority is shared in the remaining states.[44]

EPA directly implements the entire UIC program in several oil and gas producing states, including Kentucky, Michigan, New York, Pennsylvania, and Virginia.[45] **Figure 3** identifies state primacy status for the UIC program.

As noted, for Class II oil and gas related injection operations, under Section 1425, a state may be delegated primary enforcement authority without meeting EPA regulatory requirements for state UIC programs promulgated under Section 1421, provided the state demonstrates that it has an effective program that prevents underground injection that endangers drinking water sources. EPA has issued guidance for approval of state programs under Section 1425.[46] If directed by Congress to regulate hydraulic fracturing as underground injection, this regulatory approach could give states significant flexibility and thus might reduce potential regulatory costs, redundancy, and other possible impacts to the industry and the states.[47] EPA's draft guidance on the use of diesel fuels in fracturing fluids does not require revision or review of state UIC programs.

Most oil and gas producing states exercise primary enforcement authority for injection wells associated with oil and gas production (Class II wells) under SDWA Section 1425. Among these states, Alaska, California, Colorado, Indiana, Montana, and South Dakota have received primacy only for Class II wells, while EPA administers the remainder of the UIC program (Class I, III, IV, and V wells) for these states.

Table 3 lists states that regulate Class II wells under Section 1425.

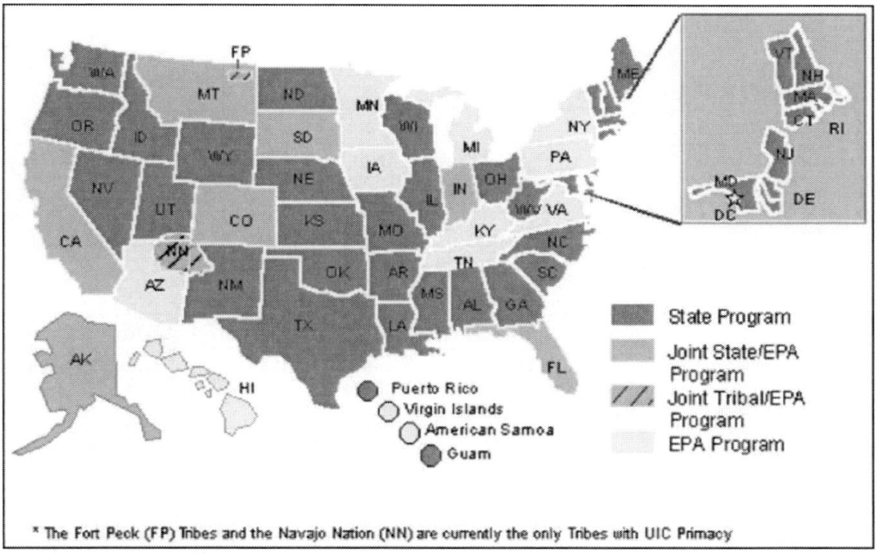

Source: U.S. Environmental Protection Agency, available at http://www.epa.gov/safewater/uic/primacy.html.

Figure 3. Primacy Status for EPA's UIC Program.

Table 3. States and Tribes Regulating Oil and Gas (Class II) UIC Wells under SDWA Section 1425

Alabama	Louisiana	Oklahoma
Alaska Arkansas	Mississippi Missouri	Oregon South Dakota
California	Montana	Texas
Colorado	Nebraska	Utah
Illinois IndianaKansas	New Mexico North Dakota Ohio	West Virginia Wyoming Assiniboine and Sioux Tribes of the Fort Peck Indian Reservation
The Navajo Nation		

Source: Adapted from information provided by U.S. Environmental Protection Agency.

Note: With primacy granted under Section 1425, states regulate Class II wells using their own program requirements rather than following EPA regulations, providing significant regulatory flexibility to the states.

Table 4. States Where EPA Implements the UIC Class II Program

Shale Gas Producing States	Others
Pennsylvania	Arizona
New York	District of Columbia
Michigan	Florida
Kentucky	Hawaii
Tennessee	Iowa
Virginia	Minnesota
	Multiple tribes, few territories

Source: U.S. Environmental Protection Agency, May 22, 2012, http://water Primacy.cfm.

Notes: Eighteen states or territories (e.g., Arizona, Maryland, New Jersey, and North Carolina) have no Class II wells. The states with the most Class II wells are Texas (52,501), California (47, 624), Kansas (15,919), and Oklahoma (10,854).

The Debate over Regulation of Hydraulic Fracturing under the SDWA

From the date of the enactment of the SDWA in 1974 until the late 1990s, hydraulic fracturing was not regulated under the act by EPA or the states tasked with administration of the SDWA. However, in the last 15 years a number of developments have called into question the extent to which hydraulic fracturing should be considered an "underground injection" to be regulated under the SDWA. A key trigger for this debate was a challenge to the Alabama UIC program brought by the Legal Environmental Assistance Foundation (LEAF).

The Legal Environmental Assistance Foundation (LEAF) Challenge to the Alabama UIC Program and EPA's Interpretation of the SDWA

In 1994, LEAF petitioned EPA to initiate proceedings to have the agency withdraw its approval of the Alabama UIC program because the program did not regulate hydraulic fracturing operations in the state associated with production of methane gas from coalbed formations.[48] The state of Alabama had previously been authorized by EPA to administer a UIC program pursuant to the terms of the SDWA.[49] EPA denied the LEAF petition in 1995 based on a finding that hydraulic fracturing did not fall within the definition of "underground injection" as the term was used in the SDWA and the EPA regulations promulgated under that act.[50] According to EPA, that term applied only to wells whose "principal function" was the placement of fluids underground.[51] LEAF challenged EPA's denial of its petition in the U.S. Court of Appeals for the Eleventh Circuit, arguing that EPA's interpretation of the terms in question was inconsistent with the language of the SDWA.[52]

The court rejected EPA's claim that the language of the SDWA allowed it to regulate only those wells whose "principal function" was the injection of fluids into the ground. EPA based this claim on what it perceived as "ambiguity" in the SDWA regarding the definition of "underground injection" as well as a perceived congressional intent to exclude wells with primarily non-injection functions.[53] The court held that there was no ambiguity in the SDWA's definition of "underground injection" as "the subsurface emplacement of fluids by well injection," noting that the words have a clear meaning and that

> The process of hydraulic fracturing obviously falls within this definition, as it involves the subsurface emplacement of fluids by forcing them into cracks in the ground through a well. Nothing in the statutory definition suggests that EPA has the authority to exclude from the reach of the regulations an activity (i.e. hydraulic fracturing) which unquestionably falls within the plain meaning of the definition, on the basis that the well that is used to achieve that activity is also used—even primarily used—for another activity (i.e. methane gas production) that does not constitute underground injection.[54]

The court therefore remanded the decision to EPA for reconsideration of LEAF's petition for withdrawal of Alabama's UIC program approval.[55]

Alabama's Regulation of Hydraulic Fracturing in CBM Production

Consideration of Alabama's UIC program after the *LEAF I* decision was issued in 1997 is a helpful case study. It is useful in assessing exactly how EPA authorized a state to regulate hydraulic fracturing under the SDWA "Class" well system, understanding the regulatory options available to EPA and the states authorized to enforce SDWA programs, and evaluating the industry impact resulting from the requirement that hydraulic fracturing be regulated under a UIC program.

Following the *LEAF I* decision and EPA's initiation of proceedings to withdraw its approval of Alabama's Class II UIC program, in 1999 Alabama submitted a revised UIC program to EPA.[56] The revised UIC program sought approval under Section 1425 of the SDWA rather than Section 1422(b). As discussed above, Section 1425 differs from Section 1422(b) in that approval under Section 1425 is based on a showing by the state that the program meets the generic requirements found in Section 1421(b)(1)(A)-(D) of the SDWA and that the program "represents an effective program (including adequate recordkeeping and reporting) to prevent underground injection which endangers drinking water sources."[57] In contrast, approval of a state program under Section 1422(b) requires a showing that the state's program satisfies the requirements of the UIC regulations promulgated by EPA.[58] In its decision on the challenge to EPA's approval of Alabama's revised UIC program, the U.S. Court of Appeals for the Eleventh Circuit observed that "the practical difference between the two statutory methods for approval is that the requirements for those programs covered under § 1425 are more flexible than the requirements for those programs covered under § 1422(b)."[59]

EPA approved Alabama's revised UIC program under Section 1425 in 2000.[60] LEAF appealed EPA's decision to the U.S. Court of Appeals for the Eleventh Circuit. LEAF made three arguments. First, LEAF claimed that EPA should not have approved state regulation of hydraulic fracturing under Section 1425 because it does not "relate to ... underground injection for the secondary or tertiary recovery of oil or natural gas," one of the requirements for approval under Section 1425.[61] The court rejected this argument, finding that the phrase "relates to" was broad and ambiguous enough to include regulation of hydraulic fracturing as being related to secondary or tertiary recovery of oil or natural gas.[62]

Second, LEAF challenged the Alabama program's regulation of hydraulic fracturing as "Class II-like" wells not subject to the same regulatory requirements as Class II wells.[63] The court agreed with LEAF on this point, noting that in its decision in *LEAF I*, it had held that methane gas production wells used for hydraulic fracturing are "wells" within the meaning of the statute.[64] As a result, the court found that wells used for hydraulic fracturing must fall under one of the five classes set forth in the EPA regulations at 40 C.F.R. Section 144.6.[65] Specifically, the court found that the injection of hydraulic fracturing fluids for recovery of coalbed methane "fit squarely within the definition of Class II wells," and as a result the court remanded the matter to EPA for a determination of whether Alabama's updated UIC program complied with the requirements for Class II wells.[66]

Finally, LEAF alleged that even if Alabama's revised UIC program was eligible for approval under Section 1425 of the SDWA, EPA's decision to approve it was "arbitrary and capricious" and therefore a violation of the Administrative Procedure Act.[67] The court rejected this argument.[68]

Among other provisions added in response to the Eleventh Circuit's decisions, the Alabama regulations prohibited fracturing "in a manner that would allow the movement of fluid containing any contaminant into a USDW, if the presence of the contaminant may (a) cause a violation of any applicable primary drinking water standard; or (b) otherwise adversely affect the health of persons."[69] The state regulations further required state approvals (but not permits) prior to individual fracturing jobs. Specifically, well operators were required to certify in writing, with supporting evidence, that a proposed hydraulic fracturing operation would not occur in a USDW, or that the mixture of fracturing fluids would meet EPA drinking water standards. Regulations also prohibited fracturing at depths shallower than 399 feet (most drinking water wells rely on shallow aquifers) and prohibited the use of diesel oil or fuel in any fracturing fluid mixture. The requirements regarding minimum depths and the diesel ban remain in place, but the rules no longer require that injection fluids meet drinking water standards. Instead, "each coal bed shall be hydraulically fractured so as not to cause irreparable damage to the coalbed methane (CBM) well, or to adversely impact any fresh water supply well or any fresh water resources."[70]

With hydraulic fracturing regulations in place, CBM development in Alabama continued. In 2009, a member of the State Oil and Gas Board of Alabama noted, "since Alabama adopted its hydraulic fracturing regulations, coalbed operators have submitted thousands of hydraulic fracturing proposals and engaged in thousands of hydraulic fracturing operations."[71]

The number of CBM well permits increased in the years following the adoption of revised regulations.[72] However, it is not clear whether, or by how much, the number of wells, the production costs, or the time required by operators may have been different without the revisions.[73] One of the requirements of the Alabama regulations in response to *LEAF I* was that fracturing fluids had to meet tap water standards where fracturing would occur within an underground source of drinking water. To ensure compliance, operators purchased water from municipal water supplies that were in compliance with federal drinking water standards to use for fracturing wells. Industry representatives have noted that if this approach were adopted for hydraulic fracturing nationwide, it would not only raise costs, but potentially put companies in competition with communities for drinking water supplies.

Some concern has been expressed that if Congress passed legislation requiring federal regulation of hydraulic fracturing broadly,[74] a separate permit might be required each time a well is hydraulically fractured, thus repeatedly disrupting oil and gas production activities. In Alabama, in response to *LEAF I*, the state did not require a permit for each fracturing operation, but rather had operators give notice and receive approval before fracturing. To further facilitate approvals for hydraulic fracturing, service companies identified to the state chemicals contained in various fracturing fluid mixtures that met the regulatory requirement that the mixtures not exceed federal drinking water standards. A well operator then could select from a list of pre-approved hydraulic fracturing fluids and provide the product name to the state, rather than have to submit separate analyses. Alabama regulations apply this approach where fracturing would occur within an underground source of drinking water.

EPA's 2004 Review of Hydraulic Fracturing for CBM Production

In response to the *LEAF I* decision, citizen reports of water well contamination attributed to hydraulic fracturing of coal beds, and the rapid growth in CBM development, EPA undertook a study to evaluate the environmental risks to underground sources of drinking water from hydraulic fracturing practices associated with CBM production. EPA issued a draft report in August 2002.[75] The draft report identified water quality and quantity problems that individuals had attributed to hydraulic fracturing of coal beds in Alabama, New Mexico, Colorado, Wyoming, Montana, Virginia, and West Virginia.[76] Based on the preliminary results of the study, EPA tentatively concluded that the potential threats to public health posed by hydraulic fracturing of coalbed methane wells appeared to be small and did not justify additional study or regulation.

EPA also reviewed whether direct injection of fracturing fluids into underground sources of drinking water posed any threat. EPA reviewed 11 major coalbed methane formations to determine whether coal seams lay within USDWs. EPA determined that 10 of the 11 producing coal basins "definitely or likely lie entirely or partially within USDWs."

In January 2003, EPA's National Drinking Water Advisory Council submitted to the EPA Administrator a report on hydraulic fracturing, underground injection control, and coalbed methane production and its impacts on water quality and water resources. The Council noted concerns regarding (1) the lack of resources to implement the UIC program, (2) the use of diesel fuel and potentially toxic additives in the hydraulic fracturing process, (3) the potential impact of coalbed methane development on local underground water resources and the quality of surface waters, and (4) the maintenance of EPA regulatory authority within the UIC program.[77]

In 2004, EPA issued a final version of the 2002 draft report, based primarily on an assessment of the available literature and extensive interviews. EPA found no confirmed cases of contamination from hydraulic fracturing of CBM formations, and concluded that the injection of hydraulic fracturing fluids into CBM wells posed little threat to underground sources of drinking water and required no further study. However, EPA found that very little documented research had been done on the environmental impacts of injecting fracturing fluids.[78] EPA had discussed the use of diesel fuel in fracturing fluids in the 2002 draft report, and concluded in the final report that "The use of diesel fuel in fracturing fluids poses the greatest potential threat to USDWs because the BTEX constituents in diesel fuel exceed the MCL [maximum contaminant level] at the point-ofinjection."[79] EPA noted that estimating the concentration of diesel fuel components and other fracturing fluids beyond the point of injection was beyond the scope of its study.[80]

In 2002, the GWPC Board of Directors passed a resolution calling for a ban on the use of diesel fuel in hydraulic fracturing of CBM wells where drinking water sources were present.[81] In 2003, EPA entered into an agreement with three companies that provided roughly 95% of hydraulic fracturing services (BJ Services, Halliburton Energy Services, and Schlumberger Technology Corporation). Under this agreement, the firms agreed to remove diesel fuel from CBM fluids injected directly into drinking water sources if cost-effective alternatives were available.[82]

EPACT 2005: A LEGISLATIVE EXEMPTION FOR HYDRAULIC FRACTURING

The decision by the U.S. Court of Appeals for the Eleventh Circuit in *LEAF I* highlighted a debate over whether the SDWA, as it read at the time, required EPA to regulate hydraulic fracturing. Although the Eleventh Circuit's decision applied only to hydraulic fracturing for coalbed methane production in Alabama, the court's reasoning—in particular, its finding that hydraulic fracturing "unquestionably falls within the plain meaning of the definition [of underground injection]"[83]—raised the issue of whether EPA could be required to regulate hydraulic fracturing under the SDWA.

Before this question was resolved through agency action or litigation, Congress passed an amendment to the SDWA as a part of EPAct 2005 (P.L. 109-58) that addressed this issue. Section 322 of EPAct 2005 amended the definition of "underground injection" in the SDWA as follows:

> The term "underground injection"—(A) means the subsurface emplacement of fluids by well injection; and (B) excludes—(i) the underground injection of natural gas for purposes of storage; and (ii) the underground injection of fluids or propping agents (other than diesel fuels) pursuant to hydraulic fracturing operations related to oil, gas, or geothermal production activities.

This amendment clarified that the UIC requirements found in the SDWA do not apply to hydraulic fracturing, although the exclusion does not extend to the use of diesel fuel in hydraulic fracturing operations. This amended language is the definition of "underground injection" found in the SDWA as of the date of this report.

EPA Guidance for Permitting Hydraulic Fracturing Using Diesel Fuels

As noted above, the EPAct 2005 amendment to the definition of "underground injection" in the SDWA excluded injections as part of hydraulic fracturing operations, but such injections involving the use of diesel fuels were not made part of the exclusion, meaning that injections for purposes of hydraulic fracturing involving the use of diesel fuel might still be made subject to regulation under the SDWA. It was not clear to states or the regulated community how EPA would address the EPAct 2005 amendment, and for several years EPA took no official position regarding the regulation of hydraulic fracturing using diesel fuels under the SDWA.[84] In 2010, EPA specified that hydraulic fracturing involving operations using diesel fuels are subject to Class II permit requirements under the SDWA, but the agency did not issue regulations or guidance to accompany this determination.

In February 2014, EPA issued final diesel permitting guidance, which states that "under the 2005 amendments to the SDWA, a UIC Class II permit must be obtained prior to conducting the underground injection of diesel fuels for hydraulic fracturing."[85] As described earlier in this report, injections subject to UIC Class II requirements must comply with a number of regulatory requirements. These include permitting requirements, and testing and monitoring obligations with respect to the well.[86] The guidance is intended for EPA permit writers and is relevant where EPA directly implements the UIC Class II program. EPA notes that "[t]o the extent that states may choose to follow some aspects of EPA guidance in implementing their own programs, it may also be relevant in areas where EPA is not the permitting authority."[87]

There had been considerable debate regarding how EPA would define "diesel fuels" in the final guidance. The draft guidance recommends using six Chemical Abstracts Service Registry Numbers (CASRNs) for determining whether diesel fuels are used in hydraulic fracturing operations.[88] These six CASRNs collectively include various types of diesel fuels, home heating oils, kerosene, crude oil, and a range of other petroleum compounds.[89] Also at issue was whether the final guidance would specify a de minimis amount of diesel fuel content for hydraulic fracturing fluids; the draft guidance did not do so. The final document covers five of the six proposed CASRNs (no longer including crude oil), and does not establish a de minimis concentration of "diesel" in fracturing fluid that would be exempt from permitting requirements.

Legislative Proposals in the 113th Congress

In the 113th Congress, several bills propose to expand federal regulation of hydraulic fracturing activities, while others would limit federal involvement. The Fracturing Responsibility and Awareness of Chemicals Act of 2013 (FRAC Act) has been introduced in the House (H.R. 1921) and the Senate (S. 1135). The bills would amend the SDWA to (1) require disclosure of the chemicals used in the fracturing process, and (2) repeal the hydraulic fracturing exemption established in EPAct 2005, and amend the term "underground injection" to include the injection of fluids used in hydraulic fracturing operations, thus authorizing EPA to regulate this process under the SDWA. Additionally, S. 1135 would authorize states to seek primary enforcement authority for hydraulic fracturing operations, regardless of whether the state had obtained primacy for other types of UIC wells, including Class II wells.

Title III of the Climate Protection Act of 2013 (S. 332) contains chemical disclosure provisions similar to the FRAC Act. S. 332 would also repeal SDWA Section 1425,[90] which provides states with an alternative to meeting the specific requirements contained in EPA UIC regulations promulgated under Section 1421 by allowing states to demonstrate to EPA that their existing programs for oil and gas injection wells are effective in preventing endangerment of underground sources of drinking water.[91] In addition, S. 332 would require EPA to report to Congress on fugitive methane emissions resulting from natural gas infrastructure.

Legislation also has been introduced to require baseline and follow-up testing of potable groundwater in the vicinity of hydraulic fracturing operations. H.R. 2983, the Safe Hydration is an American Right in Energy Development Act of 2013, would amend the SDWA to prohibit hydraulic fracturing unless the person proposing to conduct the fracturing operations agreed to testing and reporting requirements regarding underground sources of drinking water. H.R. 2983 would require testing prior to, during, and after hydraulic fracturing operations. Testing would be required for any substance EPA determines would indicate damage associated with hydraulic fracturing operations. The bill also would require EPA to post on its website all test results, searchable by zip code.

House-passed H.R. 2728, Protecting States' Rights to Promote American Energy Security Act, would amend the Mineral Leasing Act[92] to prohibit the Department of the Interior from enforcing any federal regulation, guidance, or permit requirement regarding hydraulic fracturing relating to oil, gas, or geothermal production activities on or under any land in any state that has regulations, guidance, or permit requirements for hydraulic fracturing. Although this language is broadly applicable to any federal regulation, guidance, and permit requirements "regarding hydraulic fracturing," the prohibition on enforcement applies only to the Department of the Interior, and therefore would presumably impact only hydraulic fracturing operations on lands managed by the department. The bill also would require the Department of the Interior to defer to state regulations, permitting, and guidance for all activities related to hydraulic fracturing relating to oil, gas, or geothermal production activities on federal land regardless of whether those rules were duplicative, more or less restrictive, or did not meet federal guidelines. As reported, the bill would further prohibit the department from enforcing hydraulic fracturing regulations on Trust lands, except with express tribal consent. On November 12, 2013, the House Committee on Natural Resources reported H.R. 2728, amended, and the House passed the bill, amended, on November 20. The same day, S. 1743, a companion bill to H.R. 2728, as introduced, was offered in the Senate. H.R. 2728 was placed on the Senate Legislative Calendar on December 9, 2013.

The Empower States Act of 2013, S. 1482, generally would prohibit the Secretary of the Interior from issuing regulations or guidelines regarding oil and gas production on federal land in a state if the state has otherwise met the requirements under applicable federal law. Among other provisions, the bill also would (1) amend the SDWA to require federal agencies, before issuing any oil and gas regulation or guideline, to seek comment and consult with each affected state agency and Indian tribe, and (2) require any future rule requiring disclosure of hydraulic fracturing chemicals to refer to the FracFocus database.

Other bills would limit the federal role in regulating the use of hydraulic fracturing on lands subject to federal control. The Fracturing Regulations are Effective in State Hands Act, H.R. 2513/S. 1234, would grant states sole authority to regulate hydraulic fracturing on federal lands within the state and specify that hydraulic fracturing on federal land shall be subject to the law of the state in which the land is located. This bill is likely a response to a push by the Bureau of Land Management (BLM) to adopt new regulations governing fracing on federal lands, discussed below. H.R. 1548 would prohibit the BLM hydraulic fracturing rule from having any effect on land held in trust or restricted status for Indians, except with the express consent of its Indian beneficiaries.

H.R. 2850 (H.Rept. 113-252), the EPA Hydraulic Fracturing Study Improvement Act, would require EPA to follow certain procedures governing peer review and data presentation in conducting its study on the relationship between hydraulic fracturing and drinking water. As reported, the bill would require EPA to release the final report by September 30, 2016. H.R. 2850 was placed on the Union Calendar on October 23, 2013.

Bureau of Land Management (BLM) Regulation of Hydraulic Fracturing on Federal Land

The President announced in his 2012 State of the Union address that he would require "all companies that drill for gas on public lands to disclose the chemicals they use," and in May 2012, BLM, within the Department of the Interior, proposed a rule to address the use of hydraulic fracturing in oil and gas development on public and Indian lands.[93] The proposed rule would revise

BLM oil and gas production regulations last revised in 1988,[94] and would (1) require public disclosure of chemicals used in hydraulic fracturing on BLM managed lands, (2) add new reporting and management requirements for water used in hydraulic fracturing, and (3) add new requirements related to well-bore integrity, cementing, and casing. BLM received extensive comment on the proposed rule, and in May 2013 BLM published a Supplemental Notice of Proposed Rulemaking (SNPR) and Request for Comment. BLM accepted comments through August 23, 2013, and is preparing a final rule.[95]

Changes notwithstanding, the 2012 proposed rule and the 2013 SNPR share overarching features. Both proposals would (1) add reporting and management requirements for water and other fluids used and produced in hydraulic fracturing operations, with emphasis on managing fluids that flow back to the surface, (2) require public disclosure of hydraulic fracturing chemicals, and (3) tighten well performance standards and related monitoring and reporting requirements to help ensure that wellbore integrity is maintained throughout the hydraulic fracturing process. The SNPR is narrower in that it addresses only hydraulic fracturing and excludes other well stimulation techniques. Moreover, the SNPR would require disclosure only after fracturing operations.

Potential Implications of Hydraulic Fracturing Regulation under the SDWA

The full regulation of hydraulic fracturing under the SDWA (i.e., beyond injections involving diesel) potentially could have significant, but currently unknown, environmental benefits as well as impacts on oil and natural gas producers and state and federal regulators. Resulting groundwater protection and public health benefits would likely be experienced most significantly in any states that might have relatively weaker groundwater protection provisions (such as weaker cementing and casing requirements, or allowing injection of hydraulic fracturing fluids directly into or adjacent to USDWs) compared to provisions that might be adopted by EPA. Alternatively, the possible benefits of federal regulation would likely be reduced to the degree that states currently have effective groundwater protection requirements, or respond to increased development of unconventional gas and oil resources with their own revised requirements (and numerous states have done so). The regulation of the injection of fluids for hydraulic fracturing purposes would not address surface management of chemicals or drilling wastes, or the treatment and disposal of produced water. If such surface activities were determined to be the sources of most water contamination incidents associated with unconventional oil and gas development, then federal regulation of hydraulic fracturing under the SDWA may have limited environmental and public health benefits.

Requirements for chemical disclosure are widely viewed as beneficial. The lack of information regarding chemicals used in hydraulic fracturing has made investigations of groundwater contamination difficult in some cases, because well owners and state regulators typically have not known which chemicals to test for to determine whether a fracturing compound has migrated into a water well. The debate has involved who should regulate (the states or federal government) and what should be disclosed and when. Some have called for public disclosure of chemicals in frac fluids before well stimulation so that property owners would be able to test well water for the presence of specific compounds and establish a baseline of well water quality before oil or gas development occurs. Neither the FRAC Act nor the proposed BLM rule would require chemical disclosure prior to hydraulic fracturing; however, some states (e.g., Wyoming and California) do require public disclosure prior to commencement of fracing operations.[96]

Many states have adopted a variety of disclosure requirements since the FRAC Act was first introduced. In 2011, the Ground Water Protection Council (GWPC) and Interstate Oil and Gas Compact Commission (IOGCC) established a public registry where companies may voluntarily identify chemicals used in hydraulic fracturing in specific wells. According to the GWPC, as of February 2014, at least 22 states had adopted chemical disclosure requirements, 14 states required public disclosure using FracFocus (http://www.fracfocus.org), and others were proposing to do so. **Figure 4** identifies the states that have adopted chemical disclosure requirements and the states that use, or are considering using, FracFocus.

If the SDWA were amended to authorize (but not mandate) EPA to regulate hydraulic fracturing, EPA might undertake further study to assess the potential risks of hydraulic fracturing to underground sources of drinking water. (The agency currently is conducting such studies, as discussed below.) Subsequently, EPA might determine the need for, and potential scope of, any new regulations, and decide whether to adapt the existing regulatory framework or to develop a new approach under the UIC program. The rulemaking process typically takes several years. A 2009 presentation by EPA's Region 8 UIC program explained that, if legislative change occurs,

> additional study may take place, regulations may be written by EPA, some combination of these may happen, [and] there may be a phased-in approach. If regulations are developed, they typically include: establishing a regulation development workgroup which can include the public; a proposed regulation, including opportunity for public comment (and one or more hearings if needed); a final regulation, including opportunity for judicial appeals; and an effective date for the regulation.[97]

One implication of regulating hydraulic fracturing under SDWA relates to the SDWA's citizen suit provisions. As noted, Section 1449 provides for citizen civil actions against any person or agency allegedly in violation of provisions of SDWA, or against the EPA Administrator for alleged failure to perform any action or duty that is not discretionary.[98] This provision could represent an expansion in the ability of citizens to challenge state administration of oil and gas programs related to hydraulic fracturing and drinking water, were the hydraulic fracturing exemption provision to be repealed.

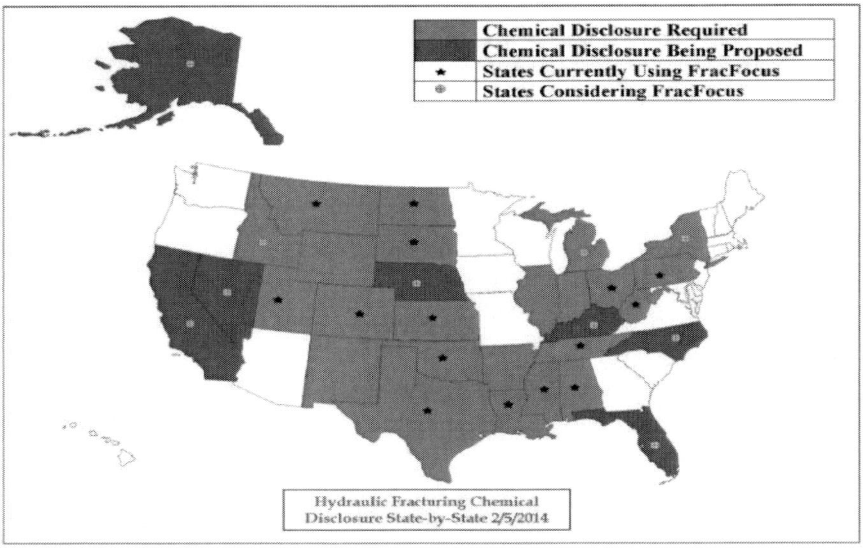

Source: Ground Water Protection Council, February 2014.

Notes: California interim rules, in place for 2014, require operators to post chemical information to the FracFocus registry, and the state also posts reports on its website. Other state actions: Alabama and North Carolina (rules being drafted).

Figure 4. Hydraulic Fracturing Chemical Disclosure by State.

As discussed, the SDWA currently includes two options for approving state UIC programs related to oil and gas recovery.[99] Under the less restrictive requirements of Section 1425, EPA may be able to implement new requirements primarily through guidance and review and approval of state programs revised to address hydraulic fracturing. EPA used this approach when ordered to require Alabama to regulate hydraulic fracturing of coal beds, and a federal district court approved this approach. For regulating the use of diesel fuel in hydraulic fracturing, EPA has drafted guidance for EPA permit writers, but has not proposed any new requirements nor has the agency proposed to review state programs.

If EPA decided to allow states to regulate hydraulic fracturing under Section 1425, the agency also might write new hydraulic fracturing regulations under Section 1421 for states such as Idaho, Maryland, and North Carolina that exercise primacy under Section 1422 (i.e., using the EPA regulations), and for EPA to use in states where EPA directly implements the UIC program (e.g., Kentucky, Michigan, New York, Pennsylvania, and Virginia). Regardless of regulatory approach, new requirements would likely require substantially more resources for UIC program administration and enforcement by the states and EPA.

The possible impacts of enacting legislation directing EPA to regulate hydraulic fracturing feasibly could vary for different oil and gas formations. The SDWA directs EPA, when developing UIC regulations, to take into consideration "varying geologic, hydrological, or historical conditions in different States and in different areas within a State."[100] Thus, if EPA were to regulate hydraulic fracturing broadly under the SDWA, the agency conceivably could establish different requirements to address such differences among states or regions. If practical and applicable, EPA might find this statutory flexibility helpful, as the USDW contamination risks of hydraulic fracturing could vary widely among different formations and settings. For example, fracturing a coal bed that may qualify as a USDW poses very different groundwater contamination risks than fracturing a shale formation that is widely separated from any USDW.[101] Thus the possible application and impact of federal regulations might vary significantly in different formations, and the impacts and potential environmental benefits would likely be greatest in formations that qualify as underground sources of drinking water or are near USDWs.[102] However, the agency has not used the flexibility in the past and might broadly apply new requirements, such as those related to well construction and cementing, and mechanical integrity testing, to protect USDWs through which wells may pass, among other purposes.

For the oil and gas industry, regulation of hydraulic fracturing under the UIC program could have a range of impacts. In some states, oil and gas operations are subject to regulation by a state oil and gas agency or commission as well as an environmental or public health agency. Industry representatives have expressed concern over the potential for some duplication of requirements from state oil and gas regulators and environmental regulators. Delays in issuing permits and commensurate delays in well stimulation and gas marketing are among the concerns. The citizen suit provision of the SDWA also may be an issue. One analysis attempting to measure the economic and energy effects of potential regulation noted that

> Experience suggests that there will be a reduction in the number of wells completed each year due to increased regulation and its impact on the additional time needed to file permits, push-back of drilling schedules due to higher costs, increased chance of litigation, injunction or other delay tactics used by opposing groups and availability of fracturing monitoring services.[103]

Several studies have attempted to estimate the potential economic and energy supply impact of regulating hydraulic fracturing under the federal UIC program. A 2009 study prepared by a consultant for DOE estimated the costs associated with "a stringent set of potential federal requirements" including (1) obtaining a permit; (2) conducting an area of review assessment; (3) performing in-situ stress analysis; (4) conducting three-dimension fracture simulation; (5) monitoring; (6) mapping fractures, or conducting other post-fracture analysis; (7) for some wells (perhaps 10%), performing state-of-the-art down-hole fracture imaging; and (8) additional cement to ensure isolation of the target zone before fracturing.[104] Based on these assumed elements of a regulatory program, the study estimated that the compliance costs for regulating hydraulic fracturing for oil and gas development would be $100,505 for new wells receiving hydraulic fracturing treatment.[105]

A stringent regulatory program under Section 1422 arguably could include many of the above requirements. However, it is unknown what EPA might require and unclear what costs would be attributed to federal regulation. Some activities already are used in the industry or required by states (e.g., well cementing across all groundwater zones).[106] EPA UIC staff note that some of the requirements assumed in the study have never been a part of the federal UIC regulations. Other effects that are not easily quantified include the costs associated with waiting periods between fracturing jobs for approvals and other potential disruptions to operations. The Ground Water Protection

Council, representing state agencies, has opposed reclassification of hydraulic fracturing as a permitted activity under the UIC programs, stating that (1) a risk has not been identified, and thus there is no evidence that [UIC] regulation is necessary; and (2) UIC regulation would divert resources from higher risk activities.[107] The legislatures of major oil and gas producing states, including the states of Alabama, Alaska, Montana, North Dakota, Wyoming, and Texas, passed and sent to Congress resolutions asking Congress not to extend SDWA jurisdiction over hydraulic fracturing activities. If authorized, EPA regulation of hydraulic fracturing under the SDWA UIC program would not address many significant public concerns often associated with the development of unconventional oil and gas resources. These concerns involve land surface disturbances associated with the development of roads, well pads, and natural gas gathering pipelines; potential impacts of water withdrawal and consumption; treatment and disposal of flowback water to surface waters; air quality impacts; noise; etc. Some of these activities are subject to other federal laws, such as Clean Water Act requirements covering the treatment and discharge of produced water into surface waters,[108] and new Clean Air Act regulations.[109] The state and federal regulatory requirements for management, treatment, and discharge of produced water may have a more significant impact on the industry than possible UIC-related requirements.[110] Other impacts related to development of unconventional oil and gas resources are highly visible and may raise more concern than the specific process of deep underground fracturing of oil and gas formations. Some of these issues (particularly land-use and facility siting issues) are beyond the reach of federal regulation, and thus are left to state and local governments to address. New York State's Revised Draft Supplemental Generic Environmental Impact Statement is one example of a state taking a comprehensive approach to addressing a broad range of possible environmental impacts that potentially could be associated with Marcellus Shale development.[111]

UIC Program Resource Issues

The funding and staffing resource implications of including hydraulic fracturing under the UIC program could be significant for regulatory agencies. Based solely on the number of wells added to the program, the workload under Class II UIC programs could more than double. Currently, there are some 172,000 Class II wells nationwide.[112] In contrast, the DOE Energy Information Administration reports that the number of producing natural gas wells in the

United States increased from 302,421 in 1999 to 514,637 wells in 2011, and that most new wells—conventional and unconventional—are fractured.[113]

EPA's annual appropriation includes funds for state grants to support state administration of many EPA programs. For the past 30 years, the annual appropriations to support state UIC programs have remained essentially flat (not accounting for inflation) at roughly $10.5 million to $11 million.[114] Ten EPA regional offices and 42 states share this amount annually to administer the full UIC program, which covers more than 700,000 wells. In 2007, the GWPC estimated that annual UIC program funding would need to increase to $56 million to fully meet the needs of the existing UIC program.[115] The GWPC further estimated that EPA would need to provide funding at a level of $100 million annually to meet the needs for the full UIC program, including the regulation of geologic sequestration of carbon dioxide. Given the large number of wells that are fractured, UIC program oversight and enforcement costs for state agencies could be considerably higher if this process was subjected to federal UIC regulations, in addition to state oil and gas rules. If authorized or directed to regulate hydraulic fracturing under the SDWA, EPA and states would need to develop new requirements for these wells and increase staff to review applications and make permitting decisions, and some integration with state oil and gas agencies would likely be needed. States and industry representatives have expressed concern that failure to provide sufficient resources would likely create permitting backlogs. For example, under UIC regulations, EPA or the primacy state must provide for a public hearing for each permit issuance, and have inspectors on site.[116] Some states impose permit fees or use other revenue-generating mechanisms, while such approaches have not been adopted in other states.

Because of the sheer number of potentially newly regulated wells, EPA (given its current resource levels) would necessarily need to rely heavily upon the states to implement this program. In 2007, the GWPC noted that states are already struggling to fully implement their UIC programs, and new requirements for hydraulic fracturing would be problematic. The GWPC cautioned that without substantial increases in funding for the UIC program,

- more states would decide to return primacy to EPA (which also would require additional funds to implement the program);
- the overall effectiveness of UIC programs would suffer as more wells and well types are added without a concurrent addition of resources to manage them;

- decisions regarding which parts of the program to fund with limited dollars could result in actual damage to USDWs if higher risk/higher cost portions of the program are put "on the back burner"; and
- negative impacts on the economy could occur as permitting times lengthen due to increased program workloads.[117]

EPA resources are also at issue. The agency would require additional technically trained staff to oversee and enforce state programs and implement the program in non-primacy states (such as Michigan, New York, and Pennsylvania). Should some states decide not to assume primacy for the new program, EPA's resource needs would grow. As with states, EPA resources are stretched. For example, the agency is continuing its review and approval of state Class V UIC programs that are being revised to implement a 1999 rulemaking. Additionally, EPA published a rule in 2010 establishing federal UIC permitting requirements for the geologic sequestration of carbon dioxide; however, no permits have been issued.

EPA HYDRAULIC FRACTURING STUDY

The use of hydraulic fracturing for oil and gas development has expanded rapidly over the past decade, and much concern has been expressed regarding the potential for this well-stimulation practice to contaminate groundwater. Although hydraulic fracturing has been applied to wells more than 1 million times in the United States[118] with little documented harm to groundwater quality, few scientific studies have been conducted to examine processes and pathways between hydraulic fracturing operations and groundwater supplies, and whether and to what extent groundwater quality was affected. A recent journal article noted the debate and uncertainty regarding the relationship between hydraulic fracturing and potable aquifers.

> Indeed many articles in newspapers, journals, and the electronic news media regarding pollution of groundwater by the hydraulic fracturing industry (e.g., Zoback et al. 2010; Molofsky et al. 2011; Osborn et al. 2011; Myers 2012; Schnoor 2012; Warner et al. 2012) convey widely differing views regarding risks of groundwater contamination by the development of unconventional gas plays. Unfortunately, little peer-reviewed scientific information is available on the hydrogeological conditions—shallow

groundwater quality in particular—associated with unconventional gas production or, for that matter, with conventional oil and gas production.[119]

In EPA's FY2010 appropriations act, Congress directed EPA to carry out a study on the relationship between hydraulic fracturing and drinking water, using a credible approach that relies on the best available science, as well as independent sources of information.[120] EPA's Hydraulic Fracturing Study Plan states that the overall purpose of the study is to understand the relationship between hydraulic fracturing and drinking water resources.[121] More specifically, EPA designed the study to examine conditions that may be associated with potential contamination of drinking water sources, and to identify factors that may lead to human exposure and risks. EPA is undertaking research studies that address the full lifecycle of water in hydraulic fracturing, from water acquisition and chemical mixing, through wastewater treatment and/or disposal.[122]

As part of the study, EPA is investigating reported incidents of drinking water contamination where hydraulic fracturing has occurred. These five retrospective case studies will be used to determine the potential relationship, if any, between reported impacts and hydraulic fracturing activities. Prospective case studies include sampling and water resource characterization before fracturing occurs, and then evaluating any water quality or chemistry changes afterward. The study's breadth and associated costs have drawn attention. The House Appropriations Committee report for the Department of the Interior, Environment, and Related Agencies Appropriation Bill, 2013 (H.Rept. 112-589), did not include the requested $4.25 million increase for additional hydraulic fracturing research. The report directed EPA to narrow the scope of study:

> Hydraulic Fracturing.—In 2010, the Committee urged EPA to research whether there is a relationship between hydraulic fracturing and drinking water. The Committee understands EPA has incorporated a review of environmental justice impacts into this study, which the Committee finds to be outside the scope of the 2010 language and an inappropriate use of funds. No funds have been provided in the bill to research environmental justice impacts related to hydraulic fracturing, and EPA shall discontinue the use of any resources that may have been diverted to this subactivity. The Committee directs the Agency to release the study's findings with respect to whether there is a relationship between hydraulic fracturing and drinking water following appropriate public comment as directed in H.Rept. 112-151 and peer review.[123] (p. 48)

In December 2012, EPA released a progress report on the hydraulic fracturing study.[124] The scope of the report covers five identified stages of the water cycle: (1) water acquisition, (2) chemical mixing, (3) well injection, (4) flowback and produced water, and (5) wastewater treatment and waste disposal, and identifies potential drinking water issues associated with each stage. The report discusses ongoing research activities that include analyzing hydraulic fracturing data collected from the oil and gas industry and states, modeling several scenarios to identify conditions that may lead to impacts on drinking water resources, conducting laboratory studies to identify impacts of discharging inadequately treated wastewater to rivers and to assess how well wastewater treatment processes remove contaminants, compiling toxicity information of chemicals, and conducting case studies. EPA notes that each research project will be peer reviewed before publication, and that

> published results from each project will be synthesized in a report of results that will inform the research questions associated with each stage of the hydraulic fracturing water cycle. The EPA has designated the report of results as a "Highly Influential Scientific Assessment," which will undergo peer review by the EPA's Science Advisory board, an independent and external federal advisory committee that conducts peer reviews of significant EPA research products and activities.... Ultimately, the results of this study are expected to inform the public and provide decision-makers at all levels with high-quality scientific knowledge that can be used in decision-making processes.[125]

The progress report does not include research results or findings.[126] EPA plans to submit a draft report of results to the Science Advisory Board for independent peer review in late 2014. A final report is expected in 2016. For FY2014, Congress provided $6.1 million for the study, and EPA has requested the same amount for FY2015.

CONCLUDING OBSERVATIONS

Hydraulic fracturing bills introduced in the 113th Congress and previously have generated considerable debate. Many state agencies have argued against regulation of hydraulic fracturing under the SDWA groundwater protection provisions, and note a long history of the successful use of this practice in developing oil and gas resources and of state regulation of the industry. Industry representatives argue that additional federal regulation would be redundant with state rules and would likely slow domestic oil and gas development and increase energy prices. At the same time, drilling and fracturing methods and technologies have changed significantly over time as they have been applied to more challenging formations, greatly increasing the amount of water, fracturing fluids, and well pressures involved in many oil and gas production operations. The increasing density of wells and geographic expansion of the use of hydraulic fracturing, along with a growing number of citizen complaints of groundwater contamination in areas undergoing oil and gas development, have led to calls for greater state and/or federal environmental oversight of this well-stimulation technique.

Central issues in the debate concern the need for, and potential benefits of, regulation of hydraulic fracturing under the SDWA. Pollution prevention generally, and groundwater protection in particular, is much less costly than cleanup, and where groundwater supplies are not readily replaceable, protection becomes a high priority. Federal regulations generally are used to address activities found to have widespread public health and environmental risks, particularly where significant regulatory gaps and unevenness exist among the states. To the extent that a regulation is needed and is well designed and implemented, public benefits (i.e., protecting underground sources of drinking water) would be expected to accrue. If Congress directed EPA to regulate fracturing under the SDWA, the environmental benefits could be significant if the risks of contamination were significant and states were not effectively addressing those risks. Alternatively, the benefits may be small if most pollution incidents were found to be related to other oil and gas production activities such as improper management of produced water or surface spills. Some of these issues are not subject to SDWA authority and would not be addressed through regulation under this act. Issues related to well construction, operation, monitoring, and closure could be addressed through the UIC program.

Thus far, the data suggest that hydraulic fracturing—particularly in deep zones—presents a low risk of contamination to underground sources of drinking water, and most reports of contamination have been associated with surface activities or well construction and operation problems, not hydraulic fracturing per se. However, while regulators and industry practitioners define hydraulic fracturing as a specific well stimulation operation, concerned individuals, the media, and others often use the term to refer broadly to the full range of activities associated with tight oil and gas production. The answer to the question "is hydraulic fracturing contaminating drinking water supplies?" may depend on how broadly one defines hydraulic fracturing.

State oil and gas and groundwater protection agencies widely support keeping responsibility for regulating hydraulic fracturing with the states. In September 2009, the GWPC—representing state groundwater protection agencies—approved a resolution supporting continued state regulation of hydraulic fracturing and encouraging Congress, EPA, DOE, and others to work with the states and the GWPC to evaluate the risks posed by hydraulic fracturing. The GWPC and others have expressed concern that regulation of hydraulic fracturing under the SDWA would divert compliance and enforcement resources from higher-priority issues. Additionally, the IOGCC—representing state oil and gas agencies—has adopted a resolution urging Congress not to remove the fracturing exemption from provisions of the SDWA, noting that the process is a temporary injection-and-recovery technique and does not fit the UIC program that EPA generally developed to address the permanent disposal of wastes.

Nonetheless, given the critical importance of good quality water supplies to homeowners, farmers, ranchers, and communities, and uneven regulation across the states, many have called for a federal solution. It could be expected that the potential impact of federal regulations on states and industry would be lessened (and provide fewer added benefits) to the degree that states currently have effective requirements or respond to increased development of unconventional gas and oil resources with their own revised requirements. In the past few years, numerous major oil and gas producing states have revised their regulations in response to changes in the industry, while other states currently are developing or considering new laws and regulations.[127]

Whether state or federal, regulations require adequate resources to be administered effectively. The sheer number of wells that rely on fracturing suggests that significant new staffing and other resources might be needed by state and federal regulators to implement and enforce any new EPA requirements on top of existing state requirements. States that have compatible requirements in place to address hydraulic fracturing might experience fewer impacts.

Debate continues over the risks that hydraulic fracturing operations may pose to underground sources of drinking water, and Congress has directed EPA to study this matter. The results of this and other studies could provide a better assessment of potential risks, and particular circumstances that may be associated with such risks, and may help inform the need for additional regulation—whether at the state level through oil and gas laws and regulations or at the federal level through the SDWA UIC program.

End Notes

[1] Hydraulic fracturing is also used for other purposes, such as developing water supply wells and geothermal production wells. This report focuses only on its use for oil and gas development.

[2] For a brief history of technological developments that have enabled unconventional gas and oil production, see U.S. Department of Energy, *Shale Gas: Applying technology to Solve America's Energy Challenges*, National Energy Technology Laboratory, March 2011, http://www.netl.doe.gov/technologies/oil-gas/publications/brochures/ Shale_Gas_March_2011.pdf.

[3] Tight gas sands are sandstone formations with very low permeability that must fractured to release the gas.

[4] According to the Schlumberger *Oilfield Glossary*, propping agents, or proppants, are "sized particles mixed with fracturing fluid to hold fractures open after a hydraulic fracturing treatment. In addition to naturally occurring sand grains, man-made or specially engineered proppants, such as resin-coated sand or high-strength ceramic materials like sintered bauxite, may also be used." The glossary is available at http://www.glossary.oilfield.slb. com/default.cfm.

[5] This process is distinct from enhanced oil and gas recovery and other secondary and tertiary hydrocarbon recovery techniques which involve separate wells. Injections for hydraulic fracturing are done through the production wells.

[6] The Schlumberger glossary notes that "produced fluid is a generic term used in a number of contexts but most commonly to describe any fluid produced from a wellbore that is not a treatment fluid. The characteristics and phase composition of a produced fluid vary and use of the term often implies an inexact or unknown composition." "Flowback" refers to "the process of allowing fluids to flow from the well following a treatment, either in preparation for a subsequent phase of treatment or in preparation for cleanup and returning the well to production."

[7] U.S. Department of Energy, Office of Fossil Energy and National Technology Laboratory, *Modern Shale Gas Development in the United States: A Primer*, DE-FG26-04NT15455, April 2009, p. 66, http://fossil.energy general/Shale_Gas_Primer_2009.pdf.

[8] Independent Petroleum Association of America, "Hydraulic Fracturing: Effects on Energy Supply, the Economy, and the Environment," fact sheet, April 2008, http://energyindepth. org/docs/pdf/Hydraulic-Fracturing-3-E%27s.pdf.

[9] American Petroleum Institute, *Hydraulic Fracturing*, http://www.api.org/oil-

[10] U.S. Energy Information Administration, *What is Shale Gas and Why is It Important?*, Energy in Brief, December 5, 2012, http://www.eia.gov/energy The U.S. Geological Survey's National Assessment of Oil and Gas Resources Update (2013) is available at http://energy.usgs.gov/OilGas/AssessmentsData/NationalOil Gas Assessment /AssessmentUpdates.aspx.

[11] U.S. Department of Energy, Office of Fossil Energy and National Technology Laboratory, *Modern Shale Gas Development in the United States: A Primer*, April 2009, pp. 47-48, http://www.netl.doe.gov/technologies/oil publications/EPreports/Shale_Gas_Primer_ 2009.pdf. Emphasis added.

[12] Using slickwater fracturing increases the rate at which fluid can be pumped down the wellbore to fracture the shale. The process may involve the use of friction reducers, biocides, surfactants, and scale inhibitors. Biocides prevent bacteria from clogging wells; surfactants

help keep the sand or other proppant suspended. Slickwater fracturing was first used in the Barnett Shale in Texas.

[13] EPA reviewed 11 major coalbed methane formations to determine whether coal seams lay within USDWs. EPA determined that 10 of the 11 producing coal basins "definitely or likely lie entirely or partially within USDWs." (U.S. Environmental Protection Agency, *Evaluation of Impacts to Underground Sources of Drinking Water by Hydraulic Fracturing of Coalbed Methane Reservoirs*, Final Report, EPA-816-04-003, Washington, DC, June 2004, p. 4-1.)

[14] The GWPC is a national association representing state groundwater and underground injection control (UIC) agencies whose mission is to promote protection and conservation of groundwater resources for beneficial uses. http://www.gwpc.org.

[15] Ground Water Protection Council, U.S. Department of Energy, Office of Fossil Energy, National Energy Technology Laboratory, *State Oil and Natural Gas Regulations Designed to Protect Water Resources*, May 2009, p. 24, http://www.gwpc.org/sites/default/files/state

[16] American Petroleum Institute, API HF1, *Hydraulic Fracturing Operations-Well Construction and Integrity Guidelines*, First Edition, October 2009, http://www.api.org/policy api_hf1_hydraulic_fracturing_operations.aspx.

[17] The scope of this report is limited to potential issues related to hydraulic fracturing and contamination of underground sources of drinking water related to the fracturing process. The management of "flowback" from the fracturing/drilling process also presents environmental and regulatory issues and also water treatment infrastructure issues. Disposal of produced water by means other than disposal through injection wells is regulated pursuant to the Clean Water Act. For a discussion of the hydraulic fracturing process and water treatment and disposal issues and regulation, see CRS Report R42333, *Marcellus Shale Gas: Development Potential and Water Management Issues and Laws*, by Mary Tiemann et al. For a discussion of Clean Water Act requirements governing discharges of pollutants, see Environmental Protection Agency (EPA), *Natural Gas Drilling in the Marcellus Shale: NPDES Program Frequently Asked Questions*, March 16, 2011, http://www.epa.gov/npdes/pubs/hydrofracturing_faq.pdf. EPA has initiated a rulemaking to regulate discharges of wastewater produced by shale gas extraction. See EPA website, *Effluent Guidelines (Clean Water Act section 304(m)): 2010 Effluent Guidelines Program Plan*, http://water.epa.gov/lawsregs/ lawsguidance/cwa/304m/.

[18] For a discussion of environmental concerns and recommendations, see, for example, Environmental Working Group, *Drilling Around the Law*, January 2010, http://static.ewg.org/files/EWG-2009drillingaroundthelaw.pdf.

[19] New York State Department of Environmental Conservation, *Fact Sheet: What We Learned from Pennsylvania*, NYS DEC NEWS, http://www.dec.ny.gov/energy

[20] For a discussion and comparison of major elements of state oil and gas rules, see, for example, Resources for the Future, RFF Center for Energy Economics and Policy, A Review of Shale Gas Regulations by State, http://www.rff.org/centers/energy. The FracFocus website (www.fracfocus.org) contains links to each state's oil and gas regulations.

[21] The Department of the Interior, Environment, and Related Agencies Appropriations Act, 2010. P.L. 111-88, H.Rept. 111-316.

[22] A 2012 Pacific Institute study found that many individuals interviewed for the study defined "hydraulic fracturing" much more broadly than the industry meaning of the term (i.e., injection of fluids into a production well). These individuals used the term broadly to include well construction, completion, and other associated activities. Noting the differences, the authors concluded that "additional work is needed to clarify terms and

definitions associated with hydraulic fracturing to support more fruitful and informed dialog and to develop appropriate energy, water, and environmental policy." See *Hydraulic Fracturing and Water Resources: Separating the Frack from the Fiction*, p. 29, http://www.pacinst.org/reports/fracking/.

[23] In the Energy Policy Act (EPAct) of 2005 (P.L. 109-58, §322), Congress amended the definition of "underground injection" in the SDWA to specifically exclude the injection of fluids or propping agents (other than diesel fuels) pursuant to hydraulic fracturing operations related to oil, gas, or geothermal production activities.

[24] *Permitting Guidance for Oil and Gas Hydraulic Fracturing Activities Using Diesel Fuels: Underground Injection Control Program Guidance #84*, Office of Water, EPA 816-R-14-001 (February 2014).

[25] The Safe Drinking Water Act of 1974 (P.L. 93-523) authorized the UIC program at EPA. UIC provisions are contained in SDWA Part C, §§1421-1426; 42 U.S.C. §§300h-300h-5.

[26] 42 U.S.C. §300h(b)(2).

[27] P.L. 109-58, §322.

[28] 42 U.S.C. §300h(d).

[29] *Id.*

[30] 42 U.S.C. §300h-1. The minimum requirements for a state UIC program can be found at 40 C.F.R. Part 145.

[31] 42 U.S.C. §300h-4. SDWA Section 1425 was added by the Safe Drinking Water Act Amendments of 1980, P.L. 96- 502. The House committee report accompanying the legislation that added Section 1425 noted that

> Most of the 32 states that regulate underground injection related to the recovery or production of oil or natural gas (or both) believe they have programs already in place that meet the minimum requirements of the Act including the prevention of underground injection which endangers drinking water sources. This is especially true of the major producing states where underground injection control programs have been underway for years. It is the Committee's intent that states should be able to continue these programs unencumbered with additional Federal requirements if they demonstrate that they meet the requirements of the Act. (U.S. House of Representatives, Committee on Interstate and Foreign Commerce, *Safe Drinking Water Act Amendments*, H. Rept. 96-1348 to accompany H.R. 8117, 96th Congress, 2d Session, September 19, 1980, p. 5.)

[32] SDWA Section 1425 requires a state to demonstrate that its UIC program meets the requirements of Section 1421(b)(1)(A) through (D) and represents an effective program (including adequate record keeping and reporting) to prevent underground injection which endangers underground sources of drinking water. To receive approval under Section 1425's optional demonstration provisions, a state program must include permitting, inspection, monitoring, and record-keeping and reporting requirements.

[33] 42 U.S.C. §300i. The Administrator may take action when information is received that (1) a contaminant is present in or is likely to enter a public drinking water supply system or underground source of drinking water "which may present an imminent and substantial endangerment to the health of persons," and (2) the appropriate state or local officials have not taken adequate action to protect such persons.

[34] 42 U.S.C. §300h(b)(1).

[35] EPA further explained this requirement in a 1993 memorandum which provided that "[t]o better quantify the definition of USDW, EPA determined that any aquifer yielding more than 1 gallon per minute can be expected to provide sufficient quantity of water to serve a

public water system and therefore falls under the definition of a USDW." EPA Memorandum: *Assistance on Compliance of 40 CFR Part 191 with Ground Water Protection Standards*. From James R. Elder, Director, Office of Ground Water and Drinking Water, to Margo T. Oge, Director, Office of Radiation and Indoor Air, June 4, 1993.

[36] §40 C.F.R. 144.3. According to EPA regulations, an exempted aquifer is an aquifer, or a portion of an aquifer, that meets the criteria for a USDW, for which protection has been waived under the UIC program. Under 40 C.F.R. Part 146.4, an aquifer may be exempted if it is not currently being used—and will not be used in the future—as a drinking water source, or it is not reasonably expected to supply a public water system due to a high total dissolved solids content. The SDWA does not mention aquifer exemption, but EPA explains that without aquifer exemptions, certain types of energy production, mining, or waste disposal into USDWs would be prohibited. EPA, typically at the Region level, makes the final determination on granting all exemptions.

[37] U.S. Environmental Protection Agency, *Evaluation of Impacts to Underground Sources of Drinking Water by Hydraulic Fracturing of Coalbed Methane Reservoirs*, EPA 816-R-04-003, June 2004, pp. 1-5.

[38] Regulatory requirements for state UIC programs are established in 40 C.F.R. §§144-147.

[39] EPA notes that state requirements "can be, and often are, more stringent than minimum federal standards." Underground Injection Control 101, Permitting Guidance for Hydraulic Fracturing Using Diesel Fuels, Technical Webinars, May 9-16, 2011.

[40] U.S. Environmental Protection Agency, Underground Injection Control Program, Classes of Wells, http://water. The inventory of Class V wells is incomplete.

[41] This approach also would parallel the agency's response to a court ruling on hydraulic fracturing (discussed below under "The Legal Environmental Assistance Foundation (LEAF) Challenge to the Alabama UIC Program and EPA's Interpretation of the SDWA.")

[42] Regulations for wells related to oil and gas production (Class II wells) are located at 40 C.F.R. Parts 144 and 146.

[43] EPA historically has differentiated Class II wells from production wells.

[44] This discussion excludes EPA's new Class VI well category for geologic sequestration of carbon dioxide.

[45] To receive primacy, a state, territory, or Indian tribe must demonstrate to EPA that its UIC program is at least as stringent as the federal standards; the state, territory, or tribal UIC requirements may be more stringent than the federal requirements. For Class II wells, states must demonstrate that their programs are effective in preventing endangerment of underground sources of drinking water (USDWs). Requirements for state UIC programs are established in 40 C.F.R. §§144-147.

[46] U.S. Environmental Protection Agency, *Guidance for State Submissions under Section 1425 of the Safe Drinking Water Act*, Ground Water Program Guidance #19, p. 20, http://www.epa.gov/safewater/uic/pdfs/guidance guide_uic_guidance-19_primacy_app.pdf.

[47] The House report for the 1980 Safe Drinking Water Act Amendments, H.R. 8117, which established Section 1425, states that "So long as the statutory requirements are met, the states are not obligated to show that their programs mirror either procedurally or substantively the Administrator's regulations." H. Report to accompany H.R. 8117, No. 96-1348, September 19, 1980, p. 5.

[48] Legal Environmental Assistance Foundation, Inc. v. U.S. Environmental Protection Agency, 118F.3d 1467, 1471 (11th Cir. 1997) ("*LEAF I*").

[49] *Id.* at 1470.

[50] *Id.* at 1471.

[51] *Id.*
[52] *Id.* at 1472,
[53] *Id.* at 1473-74.
[54] *Id.* at 1474-75.
[55] *Id.* at 1478.
[56] *See* 64 Fed. Reg. 56986 (October 22, 1999).
[57] 42 U.S.C. §300h-4(a).
[58] *Id.* at §300h-1(b)(1)(A).
[59] Legal Environmental Assistance Foundation, Inc. v. U.S. Environmental Protection Agency, 276 F.3d 1253, 1257 (11th Cir. 2001) (*LEAF II*).
[60] 65 Fed. Reg. 2889 (October 2000).
[61] *Id.* at 1256.
[62] *Id.* at 1259-61.
[63] *Id.* at 1256.
[64] *Id.* at 1262.
[65] *Id.* at 1263.
[66] *Id.* at 1263-64.
[67] *Id.* at 1256 (referring to 5 U.S.C. §706(2)(A).
[68] *Id.* at 1265.
[69] Ala. Admin. Code, r. 400-3-8-.03(4), (2002). Responding to EPAct 2005 (see below), the state made some revisions to its regulations for hydraulic fracturing of coal beds in 2007. Ala. Admin. Code r. 400-3-8-.03(1).
[70] Ala. Admin. Code r. 400-3-8-.03(1).
[71] S. Marvin Rogers, State Oil and Gas Board of Alabama and Chairman, IOGCC Legal and Regulatory Affairs Committee, *History of Litigation Concerning Hydraulic Fracturing to Produce Coalbed Methane*, January 2009, p. 5.
[72] Ala. Admin. Code r. 400-3-8-.03(6)(a), 2002. To mitigate its increased administrative costs associated with implementation of the added regulations, operators pay a fee of $175 for each coalbed group fractured.
[73] A representative of the Alabama Coalbed Methane Association noted that the costs of hydraulic fracturing are very site specific and vary with operators as well as geology.
[74] Currently, EPA has authority to regulate only the use of diesel fuel in fracturing operations.
[75] U.S. Environmental Protection Agency. Draft Evaluation of Impacts to Underground Sources of Drinking Water by Hydraulic Fracturing of Coalbed Methane Reservoirs. EOA 816-D-02-006, August 2002.
[76] *Id.*, p. 6-20-6-21.
[77] National Drinking Water Advisory Council. Report on Hydraulic Fracturing and Underground Injection Control and Coalbed Methane by the National Drinking Water Advisory Council Resulting from a Conference Call Meeting Held December 12, 2002. Washington DC.
[78] U.S. Environmental Protection Agency, *Evaluation of Impacts to Underground Sources of Drinking Water by Hydraulic Fracturing of Coalbed Methane Reservoir*s, Final Report, EPA-816-04-003, Washington, D.C., June 2004, p. 4-1. The EPA study focused specifically on CBM wells and did not review the use of hydraulic fracturing in other geologic formations, such as the Marcellus Shale or other tight oil and gas formations.
[79] *Evaluation of Impacts to USDWs by Hydraulic Fracturing of Coalbed Methane Reservoirs*, Final Report, p. 4-19.
[80] *Id.*, p. 4-12. BTEX is the acronym for benzene, toluene, ethylbenzene and xylene, which are compounds typically found in petroleum products such as gasoline and diesel fuel. These

compounds are common indicators of gasoline, diesel, or other petroleum product contamination.

[81] Ground Water Protection Council, U.S. Department of Energy, Office of Fossil Energy, National Energy Technology Laboratory, *State Oil and Natural Gas Regulations Designed to Protect Water Resources*, May 2009, p. 22.

[82] *Memorandum of Agreement Between the United States Environmental Protection Agency and BJ Services Company, Halliburton Energy Services, Inc., and Schlumberger Technology Corporation*, December 12, 2003.

[83] *LEAF I*, 118 F.3d at 1475.

[84] In January 2011, an investigation led by Representatives Waxman, Markey and DeGette of the House Committee on Energy and Commerce found that, "oil and gas service companies have injected over 32 million gallons of diesel fuel or hydraulic fracturing fluids containing diesel fuel in wells in 19 states between 2005 and 2009." http://democrats.energycommerce.house.gov/index.php?q=news/waxman-markey-and-degette-investigation-findscontinued-use-of-diesel-in-hydraulic-fracturing-f/.

[85] U.S. Environmental Protection Agency, Permitting Guidance for Oil and Gas Hydraulic Fracturing Activities Using Diesel Fuels: Underground Injection Control Program Guidance #84, EPA 816-R-14-001, February 2014, p. 1, http://water.epa.gov/type/groundwater/uic/class2/hydraulicfracturing/hydraulic-fracturing.cfm.

[86] 40 C.F.R. §124 and §§144-147.

[87] "Permitting Guidance for Oil and Gas Hydraulic Fracturing Activities Using Diesel Fuels—Draft," 77 *Federal Register* 27542.

[88] EPA explains that "diesel fuels may be used in hydraulic fracturing operations as a primary base (or carrier) fluid, or added to hydraulic fracturing fluids as a component of a chemical additive to adjust fluid properties (e.g., viscosity and lubricity) or act as a solvent to aid in the delivery of gelling agents. Some chemicals of concern often occur in diesel fuels as impurities or additives. Benzene, toluene, ethylbenzene, and xylene compounds (BTEX) are highly mobile in ground water and are regulated under national primary drinking water regulations because of the risks they pose to human health." Source: *FACT SHEET: Underground Injection Control (UIC) Program Permitting Guidance for Oil and Gas Hydraulic Fracturing Activities Using Diesel Fuels, UIC Program Guidance #84—Draft*, EPA 816-K-12-001.

[89] 77 *Federal Register* 27453. EPA explains that these CASRNs were selected "because either their primary name, or their common synonyms contained the term "diesel fuel" and they meet the chemical and physical properties "diesel fuel" as provided in the Toxic Substances Control Act (TSCA) Inventory.

[90] 42 U.S.C. §300h-4.

[91] S. 332, Section 302, would require EPA to report to Congress on fugitive methane emissions resulting from natural gas infrastructure.

[92] 30 U.S.C. §181 et seq.

[93] Department of the Interior, Bureau of Land Management, "Oil and Gas; Well Stimulation, Including Hydraulic Fracturing, on Federal and Indian Lands," 77 *Federal Register* 27691, May 11, 2012. On federal lands, BLM administers leasing and coordinates planning and permitting with other federal agencies, as appropriate. BLM received more than 170,000 comments on the 2012 proposed rule.

[94] The proposed rule would amend existing BLM regulations at 43 C.F.R. Section 316.3-2, and add new Section 3162.3-3. Existing Section 3162.3-3 would be renumbered.

[95] Department of the Interior, Bureau of Land Management, "Oil and Gas; Hydraulic Fracturing on Federal and Indian Lands: Supplemental Notice of Proposed Rulemaking," 78 *Federal Register* 31636, May 24, 2013. BLM received roughly 1,340,000 comments on this supplemental proposal.

[96] For a review of state and proposed BLM disclosure requirements, see CRS Report R42461, *Hydraulic Fracturing: Chemical Disclosure Requirements*, by Brandon J. Murrill and Adam Vann.

[97] U.S. Environmental Protection Agency, Region 8, *Hydraulic Fracturing*, Presentation, Underground Injection Control Program Meeting, Glenwood Springs, Colorado, August 8, 2009.

[98] §1449; 42 U.S.C. 300j-8.

[99] In the case concerning Alabama, the Eleventh Circuit Court of Appeals ruled that "EPA's decision to subject hydraulic fracturing to approval under § 1425 rests upon a permissible construction of the Safe Drinking Water Act." *Legal Environmental Assistance Fund v. Environmental Protection Agency, State Oil and Gas Board of Alabama*, 276 F.3d 1253 (11th Cir. 2001).

[100] §1421(b)(3)(A); 42 U.S.C. 300h(b)(3)(A).

[101] Because coal beds frequently are sources of drinking water, the Alabama State Oil and Gas Board requires well operators to certify that a proposed hydraulic fracturing operation would not occur in a USDW, or that the mixture of fracturing fluids would meet EPA drinking water standards. State rules also prohibit fracturing at depths shallower than 399 feet, as most drinking water wells rely on shallow aquifers.

[102] U.S. Department of Energy Office of Fossil Energy and National Technology Laboratory, *Modern Shale Gas Development in the United States: A Primer*, DE-FG26-04NT15455, April 2009, http://fossil.energy oilgas/publications/naturalgas_general /Shale_Gas_Primer_2009.pdf.

[103] IHS Global Insight, *Measuring the Economic and Energy Impacts of Proposals to Regulate Hydraulic Fracturing*, Task 1 Report, Prepared for the American Petroleum Institute, Lexington, MA, 2009, p. 7.

[104] Advanced Resources International, Inc., *Potential Economic and Energy Supply Impacts of Proposals to Modify Federal Environmental Laws Applicable to the U.S. Oil and Gas Exploration and Production Industry*, U.S. Department of Energy, Office of Fossil Energy, January 2009. The authors note that cost estimates are based on a 1999 memorandum prepared for DOE, from Robin Petrusak, ICF Consulting to Nancy Johnson, U.S. Department of Energy, "Documentation of Estimated Potential Cost of Compliance for Toxic Release Inventory (TRI) Reporting and Hydraulic Fracturing," August 19, 1999.

[105] *Id.*, pp. 25-26.

[106] In discussing lessons learned from developing the Barnett Shale, industry consultants reported that an "important factor, requiring 3D seismic [imaging], is the avoidance of geo-hazards, such as water-bearing karsts and faults." Scott Stevens and Vello Kuuskraa, Advanced Resources International, Inc., "Gas Shale-1: Seven Plays Dominate North America Activity," *Oil & Gas Journal*, vol. 107, no. 36 (September 28, 2009), p. 41.

[107] Statement of Scott Kell, for the Ground Water Protection Council, House Committee on Natural Resources, Subcommittee on Energy and Mineral Resources, Oversight Hearing on "Unconventional Fuels, Part I: Shale Gas Potential," June 4, 2009.

[108] EPA has initiated a rulemaking to control the discharge of wastewater produced by shale gas extraction. EPA plans to propose a rule in 2014 that would establish discharge standards for

wastewater from shale gas extraction. See EPA website, *Effluent Guidelines (Clean Water Act section 304(m)): 2010 Effluent Guidelines Program Plan*, http://water

[109] In August 2012, EPA promulgated emission standards for air pollutants from oil and gas wells and related production systems and activities. (See CRS Report R42833, *Air Quality Issues in Natural Gas Systems*, by Richard K. Lattanzio.

[110] U.S. Department of Energy, Office of Fossil Energy and National Technology Laboratory, *Modern Shale Gas Development in the United States: A Primer*, pp. 29-42.

[111] New York imposed a moratorium on unconventional gas drilling until the state can update oil and gas regulations to govern development of the Marcellus Shale and other tight shale formations in the state using hydraulic fracturing combined with directional drilling. See New York State Department of Environmental Conservation and Division of Mineral Resources, *Revised Draft Supplemental Generic Environmental Impact Statement on the Oil, Gas and Solution Mining Regulatory Program: Well Permit Issuance for Horizontal Drilling and High-Volume Hydraulic Fracturing to Develop the Marcellus Shale and Other Low-Permeability Gas Reservoirs*, September 2011, http://www.dec.ny.gov/energy/75370.html.

[112] U.S. Environmental Protection Agency, Underground Injection Control Program, Classes of Wells, http://water

[113] EIA, *Natural Gas Navigator: Number of Producing Gas Wells*, August 2012, http://tonto.eia.doe.gov/dnav/ng/ ng_prod_wells_s1_a.htm. The number of producing wells declined to 482,822 in 2012.

[114] For state UIC grants, Congress provided $10.06 million for FY2013 and $10.50 million for FY2014.

[115] Ground Water Protection Council, *Ground Water Report to the Nation: A Call to Action*, Underground Injection Control, Ch. 9, Oklahoma City, OK, 2007, http://www.gwpc.org. This estimate preceded EPA's promulgation of new UIC regulations establishing Class VI wells for geologic sequestration of carbon dioxide and EPA's determination that production wells that use diesel must receive a Class II permit.

[116] See requirements at, for example, 40 C.F.R. 144.51(m), *Requirements prior to commencing injection*. Also, 40 C.F.R. Section 124.11 provides for public comments and requests for public hearings for UIC permits. The UIC program director is required to hold a public hearing whenever he or she finds a significant degree of public interest in a draft permit (40 C.F.R. §124.12(a)). Section 124.13 states that a comment period may need to be longer than 30 days to allow commenters time to prepare and submit comments.

[117] Mike Nickolaus, Ground Water Protection Council, UIC Funding Presentation, January 23, 2007.

[118] George E. King, "Hydraulic Fracturing 101: What Every Representative, Environmentalist, Regulator, Reporter, Investor, University Researcher, Neighbor and Engineer Should Know About Estimating Frac Risk and Improving Frac Performance in Unconventional Gas and Oil Wells," The Woodlands, TX, February 6-8, 2012, p. 2, http://www.kgs.ku.edu/PRS/Fracturing/Frac_Paper_SPE_152596.pdf.

[119] R.E. Jackson, A.W. Gorody, and B. Mayer et al., "Groundwater Protection and Unconventional Gas Extraction: The Critical Need for Field-Based Hydrogeological Research," *Groundwater*, vol. 51, no. 4 (July/August 2013), p. 489.

[120] P.L. 111-88, H.Rept. 111-316: Hydraulic Fracturing Study.—The conferees urge the Agency to carry out a study on the relationship between hydraulic fracturing and drinking water, using a credible approach that relies on the best available science, as well as independent sources of information. The conferees expect the study to be conducted through a

transparent, peer-reviewed process that will ensure the validity and accuracy of the data. The Agency shall consult with other Federal agencies as well as appropriate State and interstate regulatory agencies in carrying out the study, which should be prepared in accordance with the Agency's quality assurance principles.

[121] U.S. Environmental Protection Agency, *Plan to Study the Potential Impacts of Hydraulic Fracturing on Drinking Water Sources*, Office of Research and Development, EPA/600/R-11/122, November 2011, http://www.epa.gov/ hfstudy/HF_Study__Plan_110211_FINAL_508.pdf.

[122] EPA designated the final report as a "highly influential scientific assessment" (HISA), and thus will follow the peer review planning requirements described in the Office of Management and Budget's *Information Quality Bulletin for Peer Review*, 2004, http://www.whitehouse.gov/sites/default/files/omb/memoranda/fy2005/m05-03.pdf. The Bulletin states that important scientific information must be peer reviewed by qualified specialists before being disseminated by the federal government. Also, the Bulletin applies stricter minimum requirements for the peer review of highly influential scientific assessments (a subset of influential scientific information).

[123] H.Rept. 112-151—Department of the Interior, Environment, and Related Agencies Appropriation Bill, 2012, to accompany H.R. 2584, included the following provision: Hydraulic Fracturing.—The Committee directs the Agency to submit the Final Draft of the Interim Study Results and any additional final study results of the Plan to Study the Potential Impacts of Hydraulic Fracturing on Drinking Water Resources, for Interagency Review and public comment, consistent with the processes described in Sections 2.2 [Stakeholder Input] and 2.5 [Interagency Cooperation] of the Draft Hydraulic Fracturing Study Plan released February 7, 2011.

[124] U.S. Environmental Protection Agency, *Study of the Potential Impacts of Hydraulic Fracturing on Drinking Water Resources: Progress Report*, Office of Research and Development, EPA 601/R-12/011, December 2012, http://www.epa.gov/hfstudy.

[125] U.S. Environmental Protection Agency, *Plan to Study the Potential Impacts of Hydraulic Fracturing on Drinking Water Sources*, Executive Summary, p. 4.

[126] EPA is conducting retrospective case studies at five sites to develop information about the potential impacts of hydraulic fracturing on drinking water resources under different circumstances. The case studies involve the investigation of reported drinking water contamination attributed to hydraulic fracturing operations at oil or gas production sites (i.e., the Bakken Shale in Kildeer, Dunn County, ND; the Barnett Shale in Wise County, TX; the Marcellus Shale in Bradford County, Susquehanna County, and Washington County, PA; and coalbed methane in the Raton Basin, CO). EPA has analyzed two rounds of sampling data for the case studies, and results will be posted on EPA's website.

[127] California, Colorado, Idaho, Michigan, Montana, New Mexico, North Dakota, Ohio, Pennsylvania, Texas, West Virginia, Wyoming, and other states have revised or are revising their oil and gas production rules. Common changes include new requirements for well construction and operation (cementing, casing, pressure testing), wastewater management, and chemical disclosure. Colorado's rules, for example, include a well casing program to protect groundwater, require well treatment and fracturing reporting, and require operators to notify landowners at least one week before conducting hydraulic fracturing or other operations. Colorado and North Dakota also require baseline testing of nearby wells before drilling begins. New York is comprehensively reworking its regulations, and several other states are drafting new rules. For a discussion and comparison of major elements of state oil and gas rules, see, for example, Resources for the Future, RFF Center for Energy

Economics and Policy, *A Review of Shale Gas Regulations by State*, http://www.rff.org/centers/energy

INDEX

C

F

G

H

I

J

K

L

M

N

O

P

U

V

W

Y